A First Introduction to the Finite Element Analysis
Program MSC Marc/Mentat

Andreas Öchsner · Marco Öchsner

A First Introduction to the Finite Element Analysis Program MSC Marc/Mentat

Second Edition

 Springer

Andreas Öchsner
Faculty of Mechanical Engineering
Esslingen University of Applied Sciences
Esslingen
Germany

Marco Öchsner
Griffith School of Medical Science
Griffith University (Gold Coast Campus)
Southport, QLD
Australia

ISBN 978-3-319-89120-0   ISBN 978-3-319-71915-3   (eBook)
https://doi.org/10.1007/978-3-319-71915-3

1st edition: © Springer Science+Business Media Singapore 2016
2nd edition: © Springer International Publishing AG 2018
Softcover reprint of the hardcover 2nd edition 2017

Printed on acid-free paper

This Springer imprint is published by Springer Nature
The registered company is Springer International Publishing AG
The registered company address is: Gewerbestrasse 11, 6330 Cham, Switzerland

Nichtstun ist eine der größten und verhältnismäßig leicht zu beseitigenden Dummheiten.

Franz Kafka (1883–1924)

Preface

This book is a short introduction to the general-purpose finite element program MSC Marc, which is distributed by the MSC Software Corporation. It is a specialized program for nonlinear problems (implicit solver) which is common in academia and industry. The primary goal of this book is to provide a quick introduction to the software based on simple examples. The documentation of all finite element programs nowadays contains a variety of step-by-step examples of different complexities. In addition, all software companies offer professional workshops on different topics. The intention of this book is not to compete with these professional offers and opportunities. We would rather like to focus on simple examples, often single-element problems, which can easily be related to the theory which is provided in finite element lectures. In that sense, it is rather a companion book to a classical introductory course in the finite element method.

Chapter 1 starts with some historical comments on the development of finite element softwares. Then, a short introduction to the steps of a finite element analysis as well as the graphical interface Mentat is provided. Chapter 2 introduces the simplest one-dimensional element, i.e. a rod which can only deform along its principal axis. The spatial arrangement is then treated to cover more practical problems. Chapter 3 covers simple beam elements which can deform perpendicular to their primary axis. These elements are then arranged as plane frame structures under the consideration of a generalized beam element which can elongate and deflect. Chapter 4 presents a higher order beam theory according to Timoshenko. This formulation considers the contribution of the shear force on the deflection. Chapter 5 extends the rod element to a two-dimensional plane elasticity problem. Chapter 6 introduces the two-dimensional equivalent of the simple beam, a classical plate. Chapter 7 covers three-dimensional elements in the form of hexahedrons. Chapter 8 introduces a nonlinear problem, i.e. the elastoplastic deformation of rod elements. Chapter 9 summarizes a few advanced topics which are helpful for larger simulations or parametric studies.

The instructions provided in this *second edition* of the book relate to the Marc/Mentat 2017.0.0 (64 bit) version (the first edition featured the 2014.0.0 (64 bit) version). The second edition was enriched by further examples and more explanations to facilitate the interaction with the program. The graphical interface and the command structure might be slightly different for older versions and the reader is, in that case, advised to adjust some of the given instructions. The same must be expected for future versions.

We look forward to receiving some comments and suggestions for the next edition of this introductory work.

Southport, Australia   Andreas Öchsner
November 2017   Marco Öchsner

Contents

Symbols and Abbreviations

Latin Symbols (Capital Letters)

A	Area, cross-sectional area, geometrical dimension
B	Geometrical dimension
A_s	Shear area
E	Young's modulus
EI	Bending stiffness
F	Force, yield condition
G	Shear modulus
GA	Shear stiffness
GI_p	Torsional stiffness
H	Kinematic hardening modulus
I	Second moment of area,
K	Global stiffness matrix
K^e	Elemental stiffness matrix
L	Element length
M	Moment
N	Normal force
Q	Shear force
V	Volume
X	Global Cartesian coordinate
Y	Global Cartesian coordinate
Z	Global Cartesian coordinate

Latin Symbols (Small Letters)

a	Geometrical dimension
b	Geometrical dimension

c	Geometrical dimension
f	Column matrix of loads
h	Geometrical dimension
k	Yield stress
k_s	Shear correction factor
q	Distributed load
t	Geometrical dimension
u	Displacement
w	Specific work
x	Local Cartesian coordinate
y	Local Cartesian coordinate
z	Local Cartesian coordinate

Greek Symbols (Capital Letters)

Γ	Boundary
Ω	Domain

Greek Symbols (Small Letters)

α	Angle
β	Angle
γ	Shear strain
ε	Strain
κ	Curvature, isotropic hardening parameter
λ	Consistency paramter
ν	Poisson's ratio
ρ	Density
σ	Stress
ϕ	Rotation
ϕ_f	Fiber volume fraction
φ	Rotation

Mathematical Symbols

$\times$	Multiplication sign (used where essential)
$[\ldots]$	Matrix
$[\ldots]^T$	Transpose
$\mathrm{sgn}(\ldots)$	Signum (sign) function

Indices, Superscripted

$\ldots^e$	Element
$\ldots^{el}$	Elastic

$\ldots^{elpl}$	Elastoplastic
$\ldots^{init}$	Initial
$\ldots^{pl}$	Plastic

Indices, Subscripted

$\ldots_c$	Composite, compression, cylindrical
$\ldots_{eff}$	Effective
$\ldots_f$	Fiber
$\ldots_H$	Hoop
$\ldots_L$	Longitudinal
$\ldots_m$	Matrix
$\ldots_{max}$	Maximum
$\ldots_{nom}$	Nominal
$\ldots_R$	Reaction
$\ldots_s$	Shear, Hemispherical
$\ldots_t$	Tensile

Abbreviations

1D	One-dimensional
2D	Two-dimensional
3D	Three-dimensional
BC	Boundary condition
FEM	Finite element method
SI	International system of units

Chapter 1
Introduction to Marc/Mentat

1.1 Historical Comments on the Program's Development

The development of finite element software goes back to the early 60s when Richard MacNeal and Bob Schwendler in 1963 founded the MacNeal-Schwendler Corporation (MSC). In 1965 MSC was awarded the original contract from NASA to commercialize the finite element analysis software known as Nastran (NASA Structural Analysis) [13]. Nastran is still today the workhorse for large *linear* simulations.

Marc was the first commercial *non-linear* general-purpose finite element program. The development of the Marc software is significantly connected with two scientists, namely Pedro V. Marcal and David Hibbitt. From the development of the Marc software, a competitive product called Abaqus emerged and Table 1.1 illustrates the common roots of both packages. Both packages have a similar core and are popular in research because the user can write user-subroutines in the programming language Fortran, which is the language of all classical finite element packages. This subroutine feature allows the user to replace certain modules of the core code and to implement new features such as constitutive laws or new elements [8, 9]. The overview on the 'History of Nonlinear and Rubber FEA' provided in [1] contains also some major steps of the technical development of the Marc software.

1.2 General Comments on Finite Element Analyses

A finite elements analysis involves different steps and program modules. A user normally defines the computational model to be solved in the graphical interface, the so-called pre-processor. The geometry can either be created in the pre-processor or imported from external computer-aided design programs (CAD), see Table 1.2.

Most of the commercial finite element pre-processors have specific import filters open to third-party CAD files, e.g. a prt-file in the case of Pro/ENGINEER (Pro Creo). If there is no import filter available for a specific CAD package, the importing

© Springer International Publishing AG 2018

A. Öchsner and M. Öchsner, *A First Introduction to the Finite Element Analysis Program MSC Marc/Mentat*, https://doi.org/10.1007/978-3-319-71915-3_1

Table 1.1 Some historical steps in relation to the development of MSC Marc

Year	Comment	Ref.
1964	Pedro V. Marcal received a Ph.D. in Applied Mechanics from the Imperial College of Science and Technology, University of London	[20]
...	Pedro V. Marcal worked as Lecturer at the Imperial College of Science and Technology, University of London	[11]
∼1965	A group of researchers at Brown University in Providence, Rhode Island started the development of finite element software	[30]
1967–1974	Pedro V. Marcal taught as Professor in the Division of Engineering, Brown University in Providence, Rhode Island	[11]
1971	Pedro V. Marcal founded the MARC Analysis Research Corporation in Palo Alto, California	[30]
1972	The first version of MARC was introduced	[30]
1972	David Hibbitt received a Ph.D., related to computational mechanics using the finite element method, from Brown University	[30]
1972–1977	David Hibbitt worked for the MARC Analysis Research Corporation, responsible for the development of the MARC program	[6]
1976	Dr. Bengt Karlsson who worked for the Control Data Corporation (CDC) in Sweden joined MARC	[30]
1977	Paul Sorensen received a Ph.D. from Brown University and joined General Motors. He worked previously briefly for MARC	[2, 30]
1978	David Hibbitt founded Hibbitt, Karlsson & Sorensen, Inc. (later known as ABAQUS Inc.) and began the design and development of the ABAQUS program	[6]
1999	The MacNeal-Schwendler Corporation (now MSC Software Corporation) acquired the MARC Analysis Research Corporation for about $36 million	[25]
2005	Dassault Systèmes acquired ABAQUS Inc. for about $413 million	[30]
2017	Hexagon, the Swedish provider of information technologies, acquired the MSC Software Corporation for $834 million	[23]

Table 1.2 Some common CAD packages

Name	Company	Web page
AutoCAD	Autodesk	http://www.autodesk.com/
CATIA	Dassault Systèmes	http://www.3ds.com/
Pro/ENGINEER (PTC Creo)	Parametric Technology Corporation	http://www.ptc.com
Solid Edge	Siemens PLM	http://www.plm.automation.siemens.com.
SolidWorks	Dassault Systèmes	http://www.3ds.com/

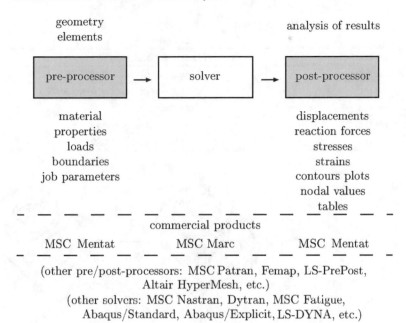

Fig. 1.1 The Marc/Mentat package

of a geometry file is still possible via a neutral file format such as ACIS, IGES, STL or STEP. It should be noted here that the generation of complex geometries is in general much easier to do in specialized CAD programs than in finite element pre-processors. Once the geometry is available in the pre-processors, the user must create and optimize the finite element mesh and assign geometrical and material properties to the elements. In a following step, the initial and boundary conditions must be defined, and the solution parameters chosen. In a next step, a text file (ASCII) is generated which contains all the information that is required to solve the problem by the solver. The solver itself is normally not visible to the user and is simply running in the background. Once the solution is obtained, the user can import the results into the so-called post-processor[1] and analyze the results. Figure 1.1 shows the general steps which are involved in a finite element analysis.

Some of the files which are created during the steps indicated in Fig. 1.1 are summarized in Table 1.3.

It should be noted here that the general steps shown in Fig. 1.1 are the main core of a finite element analysis. However, the transfer of a 'real' problem into a computational model requires much more work and engineering knowledge. In general, all 'real' problems are first idealized and assumptions must be made to simplify the problem. This refers to all areas of modelling, i.e. geometry, material, boundary conditions, among others. Depending on the question that should be answered, smaller

---

[1]The pre- and post-processor are identical in most of the finite element packages.

Table 1.3 Summary: MSC Marc/Mentat files for data storage

File name	Description
jobidname.mud	Database file, binary
jobidname.dat	Data input file for solver, ASCII
jobidname.log	Analysis sequence log file, ASCII
jobidname.out	Output file, ASCII
jobidname.t16	Post file (results), binary
jobidname.t19	Post file (results), ASCII

Fig. 1.2 Idealization of real structures

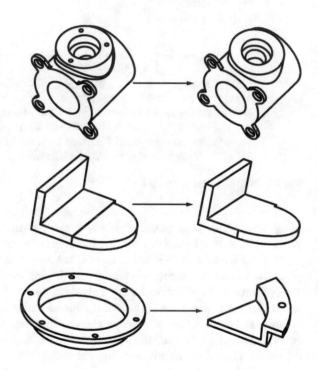

geometrical features might be disregarded and/or symmetry approaches could be used, see Fig. 1.2.

Looking at 'real' problems, it can be stated that such problems are in a strict sense three-dimensional and non-linear. Nevertheless, the discretization, i.e. the approximation of the geometry by finite elements, might be based on elements of lower dimensionality, i.e. one- or two-dimensional elements, to reduce the size of the computational model or even to make it manageable by state-of-the-art computer hardware, see Fig. 1.3. In a similar sense, the computational engineer must decide the assumptions for the material model: A pure linear-elastic material requires in general a single solution step while a non-linear material definition (e.g. plasticity) would require a much more time consuming incremental approach.

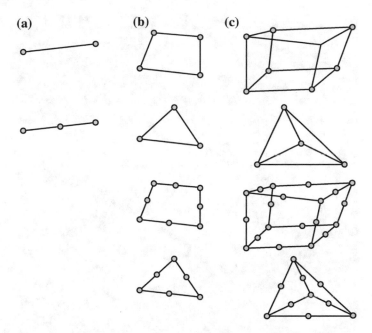

Fig. 1.3 Elements of different dimensionality: **a** one-dimensional (rods or beams), **b** two-dimensional (plane elasticity or plates), and **c** three-dimensional

Let us state at the end of this section that this practical manual should be read in conjunction with a textbook on applied mechanics [26–29] and finite element theory [5, 7, 14, 17, 22, 31, 32]. In regards to finite element 'hand calculations', the reader may refer to the following textbooks [10, 18].

1.3 Graphical User Interface

The Marc/Mentat graphical user interface has several noteworthy objects, to which we will refer by the following names:

cf. Fig. 1.4:
①: Dropdown Menu
②: Function Buttons
③: Main Menu Tabs
④: Tab Sections
⑤: Model Navigator
⑥: Graphic Interface
⑦: Graphic Interface Navigation Menu

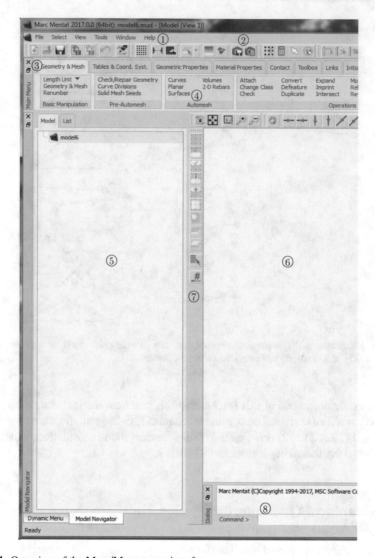

Fig. 1.4 Overview of the Marc/Mentat user interface

⑧: Command Line Dialog

cf. Fig. 1.5:
⑨: Function Dialog Windows
⑩: Buttons in Dialog Windows
⑪: Input Fields in Dialog Windows

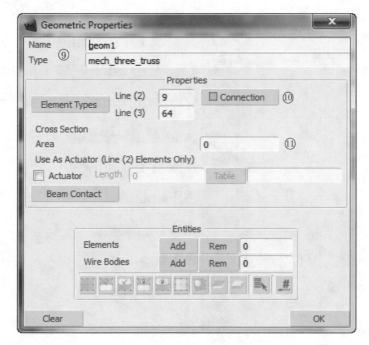

Fig. 1.5 Example of a function dialog window

When describing a certain operation in Marc/Mentat, the following notation will be used when referring to a specific object:

②: Function Buttons
③: Main Menu Tabs
④: **Tab Sections**
⑧: *Command Line Dialog*
⑨: Go to: section\subsection
⑩: Buttons in Dialog Windows
⑪: Set <variable> = <value>

Typical buttons of the graphic interface navigation menu (see ⑦ in Fig. 1.4) are shown in Fig. 1.6.

When using Marc/Mentat you will have to perform different clicks with your mouse (see Fig. 1.7):

- Left click (LC), to pick items.
- Right click (RC), to confirm selections.
- Middle click (MC), by pressing the scroll/middle button, to undo selections.

Fig. 1.6 Graphic interface navigation menu (see ⑦ in Fig. 1.4)

Select all existing

Confirm currently selected

Confirm currently unselected

Select all visible

Select invisible

All outline

All surface

All top

All bottom

Pick set

End list

1.4 Using Units

1.4.1 SI Base Units

The International System of Units (SI)[2] must be used in scientific publications to express physical units. This system consists of the seven base quantities—length, mass, time, electric current, thermodynamic temperature, amount of substance, and luminous intensity—and their respective base units are the meter, kilogram, second, ampere, kelvin, mole, and candela.[3]

---

[2]The original name is known in French as: Système International d'Unités.

[3]More information on units can be found in the brochures of the Bureau International des Poids et Mesures (BIPM): www.bipm.org/en/si.

Fig. 1.7 Schematic drawing
of a mouse

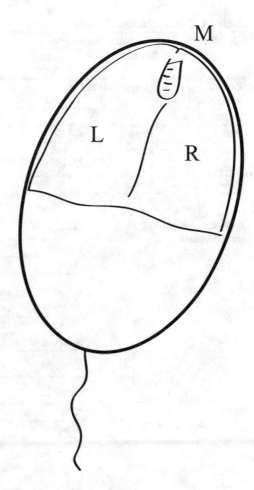

1.4.2 Coherent SI Derived Units

A coherent SI derived unit is defined uniquely as a product of powers of base units that
include no numerical factor other than 1. Table 1.4 gives some examples of derived
units and their expression in terms of base units.

1.4.3 Consistent Units

The application of a finite element code does normally not require that a specific
system of units is selected. A finite element code keeps *consistent* units throughout an
analysis and requires only that a user assigns the absolute measure without specifying
a specific unit. Thus, the units considered by the user during the pre-processing phase

Table 1.4 Example of coherent SI derived units

Quantity	Coherent derived unit			
	Name	Symbol	In terms of other SI units	In terms of SI base units
Celsius temperature	Degree Celsius	°C		K
Energy, work	Joule	J	N m	$m^2\,kg\,s^{-2}$
Force	Newton	N		$m\,kg\,s^{-2}$
Plane angle	Radian	rad	1	m/m
Power	Watt	W	J/s	$m^2\,kg\,s^{-3}$
Pressure, stress	Pascal	Pa	N/m^2	$m^{-1}\,kg\,s^{-2}$

Table 1.5 Example of consistent units

Property	Unit
Length	mm
Area	mm^2
Force	N
Pressure	$MPa = \dfrac{N}{mm^2}$
Moment	Nmm
Moment of inertia	mm^4
E-Modulus	$MPa = \dfrac{N}{mm^2}$
Density	$\dfrac{Ns^2}{mm^4}$
Time	s
Mass	$10^3 kg$

are maintained for the post-processing phase. The user must assure that the considered units are consistent, i.e. they fit each other. The following Table 1.5 shows an example of consistent units

Pay attention to the unit of density. The following example shows the conversion of the density of steel:

$$\varrho_{St} = 7.8\,\frac{kg}{dm^3} = 7.8 \times 10^3\,\frac{kg}{m^3} = 7.8 \times 10^{-6}\,\frac{kg}{mm^3}. \tag{1.1}$$

With

$$1\,N = 1\,\frac{m\,kg}{s^2} = 1 \times 10^3\,\frac{mm\,kg}{s^2} \quad und \quad 1\,kg = 1 \times 10^{-3}\,\frac{Ns^2}{mm} \tag{1.2}$$

follows the consistent density:

Property	Unit
Length	in
Area	in^2
Force	lbf
Pressure	$psi = \dfrac{lbf}{in^2}$
Moment	lbf in
Moment of inertia	in^4
E-Modulus	$psi = \dfrac{lbf}{in^2}$
Density	$\dfrac{lbf\,sec^2}{in^4}$
Time	sec

Table 1.6 Example of consistent English units

$$\varrho_{St} = 7.8 \times 10^{-9}\,\frac{Ns^2}{mm^4}. \tag{1.3}$$

Since literature reports time after time also other units, the following Table 1.6 shows an example of consistent English units:

Pay attention to the conversion of the density:

$$\varrho_{St} = 0.282\,\frac{lb}{in^3} = 0.282\,\frac{1}{in^3} \times 0.00259\,\frac{lbf\,sec^2}{in} = 0.73038 \times 10^{-3}\,\frac{lbf\,sec^2}{in^4}. \tag{1.4}$$

1.4.4 *Conversion of Important English Units to the Metric System*

The conversion between English units to the metric system is shown in Table 1.7. The example problems in the following chapter do not use a specific unit system. It can be assumed that the numbers are given in a consistent unit system and do not require any further conversion.

Table 1.7 Conversion of important U.S. customary units and British Imperial units ('English units') to metric units (m: meter; cm: centimeter; g: gram; N: newton; J: joule; W: watt)

Type	English unit	Conversion
Length	Inch	1 in $= 0.025400$ m
	Foot	1 ft $= 0.304800$ m
	Yard	1 yd $= 0.914400$ m
	Mile (statute)	1 mi $= 1609.344$ m
	Mile (nautical)	1 nm $= 1852.216$ m
Area	Square inch	1 sq in $= 1$ in$^2 = 6.45160$ cm^2
	Square foot	1 sq ft $= 1$ ft$^2 = 0.092903040$ m^2
	Square yard	1 sq yd $= 1$ yd$^2 = 0.836127360$ m^2
	Square mile	1 sq mi $= 1$ mi$^2 = 2589988.110336$ m^2
	Acre	1 ac $= 4046.856422400$ m^2
Volume	Cubic inch	1 cu in $= 1$ in$^3 = 0.000016387064$ m^3
	Cubic foot	1 cu ft $= 1$ ft$^3 = 0.028316846592$ m^3
	Cubic yard	1 cu yd $= 1$ yd$^3 = 0.764554857984$ m^3
Mass	Ounce	1 oz $= 28.349523125$ g
	Pound (mass)	1 lb$_m = 453.592370$ g
	Short ton	1 sh to $= 907184.74$ g
	Long ton	1 lg to $= 1016046.9088$ g
Force	Pound-force	1 lbf $= 1$ lb$_F = 4.448221615260500$ N
	Poundal	1 pdl $= 0.138254954376$ N
Stress	Pound-force per square inch	1 psi $= 1\ \frac{\text{lbf}}{\text{in}^2} = 6894.75729316837\ \frac{\text{N}}{\text{m}^2}$
	Pound-force per square foot	$1\ \frac{\text{lbf}}{\text{ft}^2} = 47.880258980336\ \frac{\text{N}}{\text{m}^2}$
Energy	British thermal unit	1 Btu $= 1055.056$ J
	Calorie	1 cal $= 4185.5$ J
Power	Horsepower	1 hp $= 745.699871582270$ W

Chapter 2
Rods and Trusses

2.1 Definition of Rod Elements

The definition of rod elements is summarized in Table 2.1. The derivation in lectures normally starts with the introduction of an elemental coordinate system (x) which is aligned with the principal axis of the element. Based on the definition of this element, deformations (u_{1x}, u_{2x}) can only occur along the principal axis. Assuming linear interpolation functions for the displacements, a constant elemental stress and strain is obtained.

The implementation of a rod element in a commercial finite element code is more general, i.e. based on the global coordinate system (X, Y, Z). Thus, the geometry is defined based on the global coordinates of each node (X_i, Y_i, Z_i) and the cross-sectional area A. In such a configuration, each node has three degrees of freedom, i.e. the three displacements expressed in the global coordinate system: u_{iX}, u_{iY}, u_{iZ}. Nevertheless, the stress and strain are uniaxial in the truss member. In the case of the linear straight truss (MSC Marc element type 9), the stiffness matrix is obtained based on an one-point integration rule whereas the mass matrix is obtained based on a two-point integration rule.

2.2 Basic Examples

2.2.1 1D Rod—Fixed Displacement

Problem Description

Given is a rod of length $L = 1.0$ and constant axial tensile stiffness given by $E = 20$ and an area $A = 0.5$ as shown in Fig. 2.1. At the left-hand side there is a fixed support and at the right-hand side there is a prescribed displacement of $u_0 = 0.5$. Discretize the problem with a single rod element (MSC Marc element type 9) and calculate the

© Springer International Publishing AG 2018

A. Öchsner and M. Öchsner, *A First Introduction to the Finite Element Analysis Program MSC Marc/Mentat*, https://doi.org/10.1007/978-3-319-71915-3_2

Table 2.1 Definition of rod elements

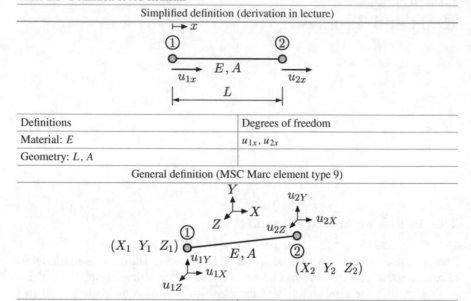

Definitions	Degrees of freedom
Material: E	u_{1x}, u_{2x}
Geometry: L, A	

General definition (MSC Marc element type 9)

Definitions	Degrees of freedom
Material: E	Node 1: u_{1X}, u_{1Y}, u_{1Z}
Geometry: A	Node 2: u_{2X}, u_{2Y}, u_{2Z}
Node 1: X_1, Y_1, Z_1	
Node 2: X_2, Y_2, Z_2	

Fig. 2.1 Schematic drawing of a single rod loaded by an end displacement u_0

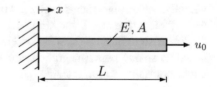

reaction force at the right-hand node.

Marc Solution

Under File → Save As..., save file as 'bar_disp'.

Constructing the mesh (Fig. 2.2)
1. Under | Geometry&Mesh |: **Basic Manipulation** select Geometry and Mesh (see Fig. 2.3).
2. Under Mesh\Nodes, select $\overline{\text{Add}}$.
3. *0,0,0* ENTER *1,0,0* ENTER.

4. Under Mesh\Elements, select $\overline{\text{Line (2)}}$ (see Fig. 2.2).
5. Press $\overline{\text{Add}}$.
6. In ⑥, select the two nodes with a (LC).
7. Press $\overline{\text{OK}}$.

Setting the Geometric Properties

8. Under $\boxed{\text{Geometric Properties}}$: **Geometric Properties** select New (Structural).
   9. Select 3D → Truss (see Fig. 2.4).
   10. Set Properties\Area = 0.5.
   11. Under Entities\Elements, press $\overline{\text{Add}}$.
12. In ⑥ select the element with a (LC), then (RC).
   13. Press $\overline{\text{OK}}$.

Fig. 2.2 Geometry and mesh dialog window

Fig. 2.3 Geometry and mesh

Setting the Material Properties

14. Under | Material Properties |: **Material Properties** select New → Finite Stiffness
Region → Standard (see Fig. 2.5).
    15. Set Other Properties\Young's Modulus = 20 (see Fig. 2.6).
    16. Under Entities\Elements, press $\overline{\text{Add}}$.

17. In ⑥ select the element with a (LC), then (RC).
    18. Press $\overline{\text{OK}}$.

Setting the Boundary Conditions

Fixed Support

19. Under | Boundary Conditions |: **Boundary Conditions** select New (Structural)
→ Fixed Displacement (see Fig. 2.7).
    20. Under Properties tick Displacement X, Displacement Y and Displacement Z.
    21. Under Entities\Nodes, press $\overline{\text{Add}}$.
22. Under ⑥ select the left most node *(0,0,0)* with a (LC), then (RC).
    23. Press $\overline{\text{OK}}$.

Fig. 2.4 Geometric
properties

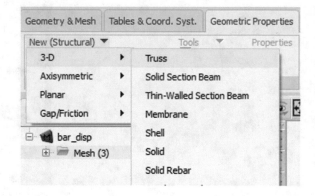

Fig. 2.5 Material properties
selection

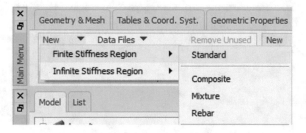

Fig. 2.6 Material properties definition

Displacement Boundary Condition

24. Under | Boundary Conditions |: **Boundary Conditions** select New (Structural) → Fixed Displacement. Set Name = disp1.

25. Under Properties tick Displacement X. Set Displacement X = 0.5 (see Fig. 2.8).

26. Under Entities\Nodes, press $\overline{\text{Add}}$.

27. Under ⑥ select the right most node *(1,0,0)* with a (LC), then (RC).

28. Press $\overline{\text{OK}}$.

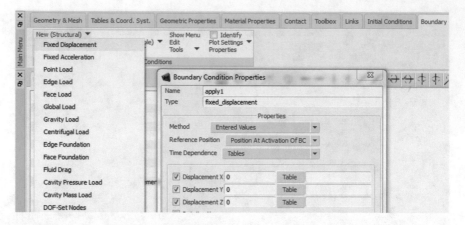

Fig. 2.7 Boundary conditions dialog window

Fig. 2.8 Entering the
displacement boundary
condition

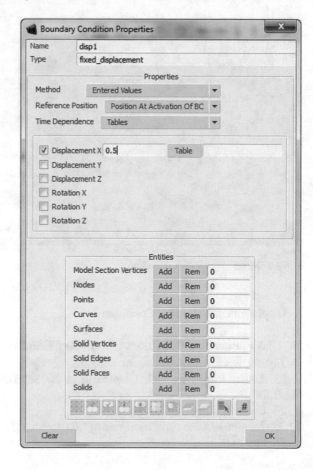

Running the Job

29. Under ⌊Jobs⌋ : **Jobs** select New → Structural.
    30. Press Check; See in ⑧ if there are any errors.
    31. If there are none, press Run.
        32. In "Run Job", press Advanced Job Submission (see Fig. 2.9).
            33. Press Save Model.
            34. Press Write Input File. Press OK.
            35. Press Submit 1.
        36. Wait until Status = Complete.
        37. Press Open Post File (Model Plot Results Menu).

Viewing the model
38. Under Deformed Shape\Style, select Deformed and Original.

Fig. 2.9 Advanced job
submission

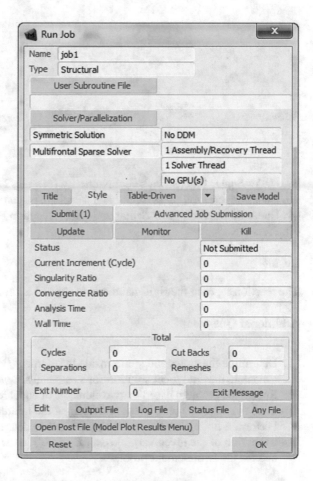

Fig. 2.10 Job result dialog
window

39. Under Scalar Plot\Style, select $\overline{\text{Numerics}}$ (see Fig. 2.10).
40. Under Scalar Plot, press $\overline{\text{Scalar}}$ and select Displacement X.
41. Under Scalar Plot, press $\overline{\text{Scalar}}$ and select Reaction Force X.
42. Press $\overline{\text{OK}}$.

Result

The reaction force at the right-hand end of the rod is found to be 5.

Additional Questions

1. Determine the stress and strain inside the element (via nodal values).
2. Calculate the analytical solution for stress and strain based on HOOKE's law.
3. Repeat the problem based on two elements of equal length ($L' = 0.5$) and determine the displacement, stress and strain at the middle and right-hand node.
4. Perform a finite element 'hand calculation' to determine the displacements (as a function of the given variables u_0, E, A, L') in the middle and the right-hand end for the two-element problem.

2.2.2 1D Rod—Fixed Point Load

Problem Description

Given is a rod of length $L = 1.0$ with a constant axial tensile stiffness given by $E = 20$ and $A = 0.5$ as shown in Fig. 2.11. At the left-hand side there is a fixed support and the right-hand side is loaded by a single force $F_0 = 5$. Use a single rod element to determine the elongation of the right-hand end.

Marc Solution

The steps for this example are the same as the one of the previous example, Sect. 2.2.1 except for steps 24–28 and the file name, which should be 'bar_force' (see Sect. 2.2.1). The following steps have to replace these:

Setting Fixed Point Load
24. Under | Boundary Conditions |: **Boundary Conditions** select New (Structural) → Point Load. Set Name = force1.
   25. Under Properties, tick Force X. Set at 5.
   26. Under Entities\Nodes, press $\overline{\text{Add}}$ (see Fig. 2.12).
27. Under ⑥ select the right most node (*1,0,0*) with a (LC), then (RC).
   28. Press $\overline{\text{OK}}$.

Result

The resulting displacement on the right-hand end of the rod is found to be 0.5.

Additional Questions

1. Determine the stress and strain inside the element (via nodal values).
2. Calculate the analytical solution for stress and strain based on HOOKE's law.
3. Repeat the problem based on two elements of equal length ($L' = 0.5$) and determine the displacement, stress and strain at the middle and right-hand node.
4. Perform a finite element 'hand calculation' to determine the displacements (as a function of the given variables u_0, E, A, L') in the middle and the right-hand end for the two-element problem.

Fig. 2.11 Schematic drawing of a single rod loaded by an end load F_0

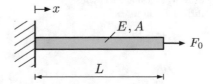

Fig. 2.12 Entering fixed
point load value

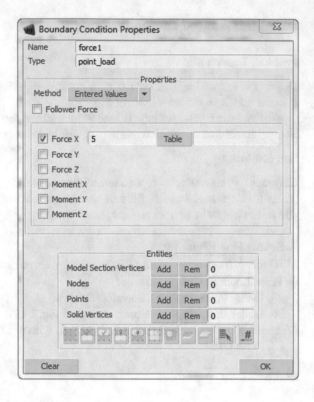

2.2.3 1D Rod—Multiple Loadcases

Problem Description

Given is a rod of length $L = 1.0$ with a constant axial tensile stiffness given by
$E = 20$ and $A = 0.5$ as shown in Fig. 2.13. At the left-hand side there is a fixed
support and the right-hand side is

(I) elongated by a given displacement $u_0 = 0.5$, or,
(II) loaded by a single force $F_0 = 5$.

Fig. 2.13 Schematic
drawing of a single rod with
different load cases

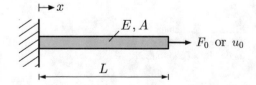

Discretize the problem with a single rod element to determine:
(a) the reaction force and
(b) the displacement at the right-hand end.

Marc Solution
For this example, involving loadcases, the steps are equal to those in example 1, from steps 1–28 (see Sect. 2.2.1). Save as 'bar_twoloads'.
Add the following steps:

Setting the Additional Boundary Condition

Force Boundary Condition

29. Under | Boundary Conditions |: **Boundary Conditions** select New (Structural) → Point Load. Set Name = force1.
    30. Under Properties, tick Force X. Set at 5.
    31. Under Entities\Nodes, press Add.
32. Under ⑥ select the right most node *(1,0,0)* with a (LC), and confirm with a (RC).
    33. Press OK.

Defining Loadcase 1—Fixed Displacement

34. Under | Loadcases |: **Loadcases** select New → Static.
    35. Set Name = fixed_displacement (see Fig. 2.14).
    36. Press Loads. Untick force1. Press OK.
    37. Set Stepping Procedure\#Steps = 1. Press OK.

Defining Loadcase 2—Fixed Point Load

38. Under | Loadcases |: **Loadcases** select New → Static.
    39. Set Name = fixed_force.
    40. Press Loads. Untick displ1. Press OK.
    41. Set Stepping Procedure\#Steps = 1 Press OK.

Running the two jobs

42. Under | Jobs |: **Jobs** select New → Structural. Set Name = force (see Fig. 2.15).
    43. Under Available, select fixed_force.
    44. Press Initial Loads. Untick displ1. Press OK.
    45. Press Check; See in ⑧ if there are any errors.

Fig. 2.14 Defining
loadcase 1

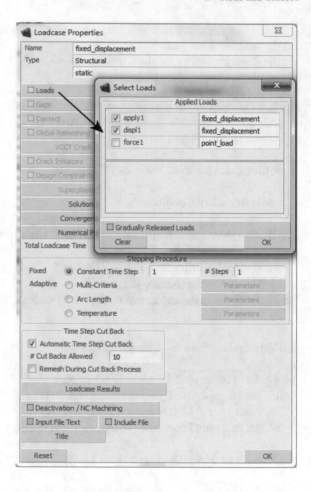

46. If there are none, press $\overline{\text{Run}}$.
    47. In 'Run Job', press $\overline{\text{Advanced Job Submission}}$.
        48. Press $\overline{\text{Save Model}}$.
        49. Press $\overline{\text{Write Input File}}$. Press $\overline{\text{OK}}$.
        50. Press $\overline{\text{Submit 1}}$.
        51. Wait until Status = Complete. Press $\overline{\text{OK}}$.
    52. Press $\overline{\text{OK}}$.

53. Under $\boxed{\text{Jobs}}$: **Jobs** select New → Structural. Set Name = displacement.
    54. Under Available, select fixed_displacement.
    55. Press $\overline{\text{Initial Loads}}$. Untick force1. Press $\overline{\text{OK}}$.
    56. Press $\overline{\text{Check}}$; See in ⑧ if there are any errors.

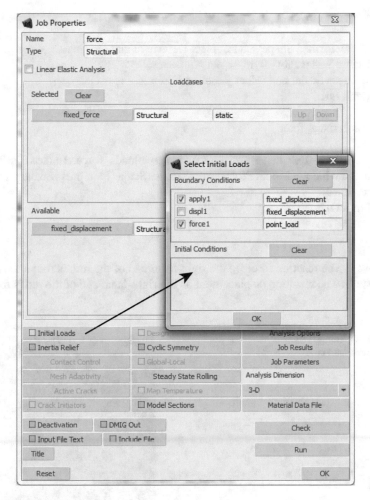

Fig. 2.15 Running Job 1

57. If there are none, press Run.
    58. In 'Run Job', press Advanced Job Submission.
        59. Press Save Model.
        60. Press Write Input File. Press OK.
        61. Press Submit 1.
    62. Wait until Status = Complete.
    63. Press Open Post File (Model Plot Results Menu).

Viewing the model

64. Under Deformed Shape\Style, select $\overline{\text{Deformed and Original}}$.
    65. Under Scalar Plot\Style, select $\overline{\text{Numerics}}$.
       66. Press $\overline{\text{Scalar}}$ and select Reaction Force X. Press $\overline{\text{OK}}$.
    67. Press $\overline{\text{OK}}$.

Open Job fixed_force

68. Under File → Results → Open. Open bar_twoloads_force.t16 (see Fig. 2.16).
    69. In the window 'Model Plot Results', under Scalar Plot, press $\overline{\text{Scalar}}$ and select Displacement X.
    70. Press $\overline{\text{OK}}$.

Results

Loadcase I: The reaction force on the right-hand end of the rod, is found to be 5.
Loadcase II: The resulting displacement on the right-hand end of the rod is found to be 0.5.

Fig. 2.16 Opening
fixed_force result case

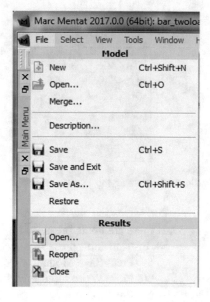

2.2.4 Plane Truss—Triangle

Problem Description

Given is the two-dimensional truss structure as shown in Fig. 2.17 where the trusses are arranged in the form of an equilateral triangle (internal angles $\beta = 60°$). The three trusses have the same length $L = 1.0$, the same YOUNG's modulus $E = 20$ and the same cross-sectional area $A = 0.5$. The structure is loaded by

(I) a horizontal force $F = 0.1$ at node 2, or,
(II) a prescribed displacement $u = 0.01$ at node 2.

Determine for (I) the deformation at the load application point and for (II) the reaction force by discretizing the structure with three rod elements.

Marc Solution

Save as 'bar_triang'

Constructing the mesh

1. Under Geometry&Mesh : **Basic Manipulation** select Geometry and Mesh.
   2. Under Mesh\Nodes, select Add.
3. *0,0,0* ENTER *0,1,0* ENTER −*0.866,0.5,0* ENTER.
   4. Under Mesh\Elements, select Line (2).
   5. Press Add.
6. In ⑥ (RC) on node 1 and 2, node 2 and 3, and node 3 and 1.
   7. Press OK.

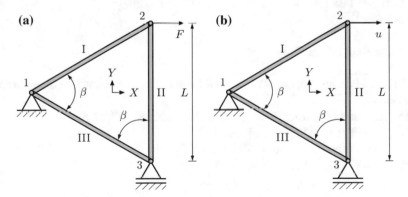

Fig. 2.17 Plane truss triangle structure: **a** force boundary condition; **b** displacement ($u \neq 0$) boundary condition

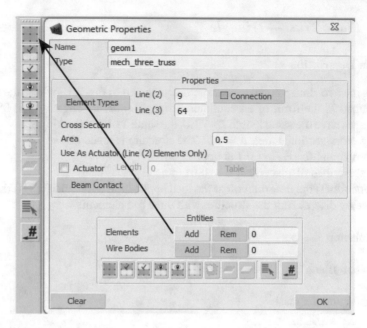

Fig. 2.18 Selecting all elements

Setting the Geometric Properties

8. Under ⎢Geometric Properties⎢: **Geometric Properties** select New (Structural).
   9. Select 3D → Truss (see Fig. 2.4).
   10. Set Properties\Area = 0.5 (see Fig. 2.18).
   11. Under Entities\Elements, press $\overline{\text{Add}}$.
12. In ⑦ (see Fig. 1.4), press the first symbol 'All Existing'.
   13. Press $\overline{\text{OK}}$.

Setting the Material Properties

14. Under ⎢Material Properties⎢: **Material Properties** select New → Finite Stiffness → Standard.
   15. Set Other Properties\Young's Modulus = 20 (See Fig. 2.6).
   16. Under Entities\Elements, press $\overline{\text{Add}}$.
17. In ⑦, press the first symbol 'All Existing'.
   18. Press $\overline{\text{OK}}$.

Setting the Boundary Conditions

Fixed Support

19. Under Boundary Conditions : **Boundary Conditions** select New (Structural)
→ Fixed Displacement (see Fig. 2.7).
   20. Under Properties tick Displacement X and Displacement Y.
   21. Under Entities\Nodes, press Add.
22. Under ⑥ select the left most node (*-0.866,0.5,0*) with a (LC), then (RC).
   23. Press OK.

24. Under Boundary Conditions : **Boundary Conditions** select New (Structural)
→ Fixed Displacement.
   25. Under Properties tick Y.
   26. Under Entities\Nodes, press Add.
27. Under ⑥ select the bottom right node (*0,0,0*) with a (LC), then (RC).
   28. Press OK.

Fixed displacement

29. Under Boundary Conditions : **Boundary Conditions** select New (Structural)
→ Fixed Displacement. Set Name = displ1.
   30. Under Properties tick Displacement X. Set Displacement X = 0.01.
   31. Under Entities\Nodes, press Add.
32. Under ⑥ select the top right node (*0,1,0*) with a (LC), then (RC).
   33. Press OK.

Point load

34. Under Boundary Conditions : **Boundary Conditions** select New (Structural)
→ Point Load. Set Name = force1.
   35. Under Properties, tick Force X. Set at 0.1.
   36. Under Entities\Nodes, press Add.
37. Under ⑥ select the top right node (*0,1,0*) with a (LC), then (RC).
   38. Press OK.

Then follow steps 34 onwards from the previous example '1D—Loadcases'
(Sect. 2.2.3).

Results

The results for the two load cases are obtained as:

(I) $u_X = 0.016$,
(II) Reaction force: $F_X = 0.059$.

Additional Questions

1. Determine the stress in each rod element.
2. Write a procedure file to automatically create the nodes and elements.
3. Perform a finite element 'hand calculation' and determine the reactions at the 'load' application point and the stresses in each rod as a function of E, L, A, F, and u. Simplify these results to the numerical values and compare with the results obtained with MSC Marc.

2.3   Advanced Examples

2.3.1   Plane Bridge Structure

Problem Description

Given is a simplified plane bridge structure over a valley as shown in Fig. 2.19. The bridge structure is idealized in the X-Y plane based on thirteen rod elements $(I, \ldots, XIII)$ which are connected at eight nodes $(1, \ldots, 8)$. Consider the following numerical values for the geometrical and material parameters: $L = 4000$ mm; $E = 200000$ MPa; and $A = 10$ mm^2. The structure is loaded by:
(I) a vertical force $F_Y = -100$ at node 2,
(II) a vertical force $F_Y = -100$ at node 3.
All elements are rod elements. Determine the deformation (u_{3X}, u_{3Y}) at node 3.

Marc Solution

Save as 'truss_bridge'

  Constructing the mesh

1. Under | Geometry&Mesh | : **Basic Manipulation** select Geometry and Mesh.
  2. Under Mesh\Nodes, select $\overline{\text{Add}}$. Add nodes 1–8.
3. (1) *0,0,0* ENTER (2) *4000,0,0* ENTER (3) *8000,0,0* ENTER (4) *12000,0,0* ENTER (5) *16000,0,0* ENTER (6) *12000,4000,0* ENTER (7) *8000,4000,0* ENTER (8) *4000,4000,0* ENTER.
  4. Under Mesh\Elements, select $\overline{\text{Line (2)}}$.
  5. Press $\overline{\text{Add}}$.
  6. In ⑥ (RC) on the following nodes to form rod elements:
       (1,2),(2,3),(3,4),(4,5),(5,6),(6,7),(7,8),
       (8,1),(8,2),(8,3),(6,3),(7,3),(6,4).
  7. Press $\overline{\text{OK}}$.

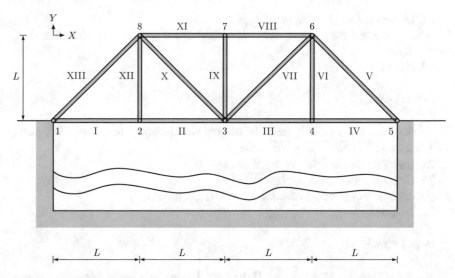

Fig. 2.19 Plane bridge structure

Setting the Geometric Properties

8. Under ⎯Geometric Properties⎯: **Geometric Properties** select New (Structural).
   9. Select 3D → Truss (see Fig. 2.4).
   10. Set Properties\Area = 10.
   11. Under Entities\Elements, press $\overline{\text{Add}}$.
12. In ⑦, press the first symbol, 'All Existing'.
   13. Press $\overline{\text{OK}}$.

Setting the Material Properties

14. Under ⎯Material Properties⎯: **Material Properties** select New → Finite Stiffness
→ Standard.
   15. Set Other Properties\Young's Modulus = 200000 (see Fig. 2.6).
   16. Under Entities\Elements, press $\overline{\text{Add}}$.
17. In ⑦, press the first symbol, 'All Existing'.
   18. Press $\overline{\text{OK}}$.

Setting the Boundary Conditions

Support Conditions

19. Under ⎯Boundary Conditions⎯: **Boundary Conditions** select New (Structural)
→ Fixed Displacement (see Fig. 2.7).
   20. Under Properties tick Displacement X and Displacement Y.

21. Under Entities\Nodes, press $\overline{\text{Add}}$.
22. Under ⑥ select the left and right most node (*0,0,0* and *16000,0,0*) with a (LC), then (RC).
23. Press $\overline{\text{OK}}$.

Point Load 1

24. Under ⎟ Boundary Conditions ⎟: **Boundary Conditions** select New (Structural) → Point Load. Set Name = force_node_2.
    25. Under Properties, tick Force Y. Set at −100.
    26. Under Entities\Nodes, press $\overline{\text{Add}}$.
27. Under ⑥ select node 2 (*4000,0,0*) with a (LC), then (RC).
    28. Press $\overline{\text{OK}}$.

Point Load 2

29. Under ⎟ Boundary Conditions ⎟: **Boundary Conditions** select New (Structural) → Point Load. Set Name = force_node_3.
    30. Under Properties, tick Force Y. Set at −100.
    31. Under Entities\Nodes, press $\overline{\text{Add}}$.
32. Under ⑥ select node 3 (*8000,0,0*) with a (LC), then (RC).
    33. Press $\overline{\text{OK}}$.

Defining Loadcase 1—Fixed force on node 2

34. Under ⎟ Loadcases ⎟: **Loadcases** select New → Static.
    35. Set Name = fixed_force_ode_2.
    36. Press $\overline{\text{Loads}}$. Untick force_node_3. Press $\overline{\text{OK}}$.
    37. Set Stepping Procedure\#Steps = 1. Press $\overline{\text{OK}}$.

Defining Loadcase 2—Fixed force on node 3

38. Under ⎟ Loadcases ⎟: **Loadcases** select New → Static.
    39. Set Name = fixed_force_node_3.
    40. Press $\overline{\text{Loads}}$. Untick force_node_2. Press $\overline{\text{OK}}$.
    41. Set Stepping Procedure\#Steps = 1 Press $\overline{\text{OK}}$.

Running the two jobs

Job 1

42. Under ⎟ Jobs ⎟: **Job** select Structural. Set Name = force_node_2.
    43. Under Available, select fixed_force_node_2.
    44. Press $\overline{\text{Initial Loads}}$. Untick force_node_3. Press $\overline{\text{OK}}$.

45. Press $\overline{\text{Check}}$; See in ⑧ if there are any errors.
46. If there are none, press $\overline{\text{Run}}$.
    47. In 'Run Job', press $\overline{\text{Advanced Job Submission}}$.
        48. Press $\overline{\text{Save Model}}$.
        49. Press $\overline{\text{Write Input File}}$. Press $\overline{\text{OK}}$.
        50. Press $\overline{\text{Submit 1}}$.
    51. Wait until Status = Complete. Press $\overline{\text{OK}}$.
52. Press $\overline{\text{OK}}$.

Job 2

53. Under $\boxed{\text{Jobs}}$: **Job** select Structural. Set Name = force_node_3.
    54. Under Available, select fixed_force_node_3.
    55. Press $\overline{\text{Initial Loads}}$. Untick force_node_2. Press $\overline{\text{OK}}$.
    56. Press $\overline{\text{Check}}$; See in ⑧ if there are any errors.
    57. If there are none, press $\overline{\text{Run}}$.
        58. In 'Run Job', press $\overline{\text{Advanced Job Submission}}$.
        59. Press $\overline{\text{Save Model}}$.
        60. Press $\overline{\text{Write Input File}}$. Press $\overline{\text{OK}}$.
        61. Press $\overline{\text{Submit 1}}$.
    62. Wait until Status = Complete.
    63. Press $\overline{\text{Open Post File}}$ (Model Plot Results Menu).

Viewing the model

64. Under Deformed Shape\Style, select $\overline{\text{Deformed and Original}}$.
    65. Under Scalar Plot\Style, select $\overline{\text{Numerics}}$.
        66. Press $\overline{\text{Scalar}}$ and select Displacement Y (or Displacement X).
        Press $\overline{\text{OK}}$.
67. Press $\overline{\text{OK}}$.

Open Job truss_bridge_force_node_2

68. Under File → Results → Open. Open truss_bridge_force_node_2.t16
    69. In the window 'Model Plot Results', under Scalar Plot\Scalar, select Displacement Y (or Displacement X).
    70. Press $\overline{\text{OK}}$.

Results

The vertical deformation of the structure is obtained for the two load cases as:

(I) $u_Y = -0.48$
(II) $u_Y = -0.97$

2.3.2 *Transmission Tower Structure*

Problem Description

Given is a simplified, i.e. plane, transmission tower structure. The transmission tower structure is idealized in the X–Y plane based on 177 rod elements which are connected by 87 nodes. Consider the following numerical values for the material and geometrical parameters: $E = 200000$ and $A = 10$. All rod elements are connected as shown in Fig. 2.20.

The load resulting from the transmission cables is approximated by vertical forces of $F_Y = -100$. Determine the deformation at the load application points.

Marc Solution

Save as 'truss_tower'

Fig. 2.20 Idealized
transmission tower structure

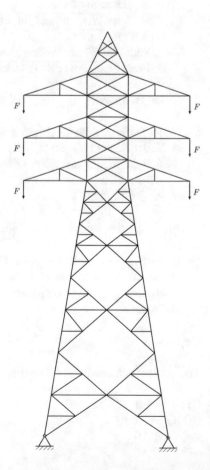

Constructing the mesh

1. Under $\boxed{\text{Geometry\&Mesh}}$: **Basic Manipulation** select Geometry and Mesh.

  2. Under Mesh\Nodes, select $\overline{\text{Add}}$. Add the following nodes:

Node Nr.	Node coordinates	Node Nr.	Node coordinates
1	(6.000, 0.000, 0)	45	(5.330, 4.000, 0)
2	(2.000, 12.00, 0)	46	(4.750, 7.500, 0)
3	(-1.00, 12.00, 0)	47	(4.330, 10.00, 0)
4	(4.000, 12.00, 0)	48	(0.330, 2.000, 0)
5	(7.000, 12.00, 0)	49	(1.000, 6.000, 0)
6	(0.700, 12.00, 0)	50	(1.500, 9.000, 0)
7	(5.300, 12.00, 0)	51	(2.330, 9.500, 0)
8	(2.000, 13.00, 0)	52	(2.125, 8.250, 0)
9	(4.000, 13.00, 0)	53	(2.125, 6.750, 0)
10	(2.000, 14.00, 0)	54	(1.830, 5.000, 0)
11	(4.000, 14.00, 0)	55	(1.830, 3.000, 0)
12	(-1.00, 14.00, 0)	56	(1.500, 1.000, 0)
13	(7.000, 14.00, 0)	57	(1.840, 11.00, 0)
14	(0.700, 14.00, 0)	58	(2.500, 11.50, 0)
15	(5.300, 14.00, 0)	59	(2.330, 10.50, 0)
16	(2.000, 15.00, 0)	60	(1.920, 11.50, 0)
17	(4.000, 15.00, 0)	61	(1.750, 10.50, 0)
18	(2.000, 16.00, 0)	62	(1.580, 9.500, 0)
19	(4.000, 16.00, 0)	63	(1.375, 8.250, 0)
20	(0.700, 16.00, 0)	64	(1.125, 6.750, 0)
21	(5.300, 16.00, 0)	65	(0.830, 5.000, 0)
22	(2.000, 17.00, 0)	66	(0.495, 3.000, 0)
23	(4.000, 17.00, 0)	67	(0.165, 1.000, 0)
24	(3.000, 19.00, 0)	68	(5.660, 2.000, 0)
25	(2.500, 18.00, 0)	69	(5.000, 6.000, 0)
26	(3.500, 18.00, 0)	70	(4.500, 9.000, 0)
27	(2.750, 18.50, 0)	71	(4.166, 11.00, 0)
28	(3.250, 18.50, 0)	72	(4.165, 3.000, 0)
29	(-1.00, 16.00, 0)	73	(4.500, 1.000, 0)
30	(7.000, 16.00, 0)	74	(4.165, 5.000, 0)
31	(0.700, 12.57, 0)	75	(3.875, 6.750, 0)
32	(5.300, 12.57, 0)	76	(3.875, 8.250, 0)
33	(0.700, 14.57, 0)	77	(3.665, 9.500, 0)
34	(5.300, 14.57, 0)	78	(3.665, 10.50, 0)
35	(0.700, 16.57, 0)	79	(3.500, 11.50, 0)
36	(5.300, 16.57, 0)	80	(4.080, 11.50, 0)
37	(0.000, 0.000, 0)	81	(4.248, 10.50, 0)
38	(3.000, 11.00, 0)	82	(4.415, 9.500, 0)
39	(3.000, 9.000, 0)	83	(4.625, 8.250, 0)
40	(3.000, 2.000, 0)	84	(4.875, 6.750, 0)
41	(3.000, 6.000, 0)	85	(5.165, 5.000, 0)
42	(0.660, 4.000, 0)	86	(5.495, 3.000, 0)
43	(1.250, 7.500, 0)	87	(5.830, 1.000, 0)
44	(1.660, 10.00, 0)		

3. Under Mesh\Elements, select $\overline{\text{Line (2)}}$.
4. Press $\overline{\text{Add}}$.
5. In ⑥ connect the nodes as shown in Fig. 2.21.
6. Press $\overline{\text{OK}}$.

Alternatively, one can also generate the model by using a procedure file (in this case 'trans_tower.proc'), a text file containing the node coordinates and the various element configurations, see Sect. 9.1 for further details.

Setting the Geometric Properties

7. Under $\boxed{\text{Geometric Properties}}$: **Geometric Properties** select New (Structural).
   8. Select 3D → Truss (see Fig. 2.4).
   9 Set Properties\Area = 10.
   10. Under Entities\Elements, press $\overline{\text{Add}}$.
11. In ⑦, press the first symbol 'All Existing'.
   12. Press $\overline{\text{OK}}$.

Setting the Material Properties

13. Under $\boxed{\text{Material Properties}}$: **Material Properties** select New → Finite Stiffness → Standard.
   14. Set Other Properties\Young's Modulus = 200000 (see Fig. 2.6).
   15. Under Entities\Elements, press $\overline{\text{Add}}$.
16. In ⑦, press the first symbol 'All Existing'.
   17. Press $\overline{\text{OK}}$.

Setting the Boundary Conditions

Boundary Supports

18. Under $\boxed{\text{Boundary Conditions}}$: **Boundary Conditions** select New (Structural) → Fixed Displacement (see Fig. 2.7).
   19. Under Properties tick Displacement X and Displacement Y.
   20. Under Entities\Nodes, press $\overline{\text{Add}}$.
21. Under ⑥ select the bottom left and right node (*0,0,0* and *6,0,0*) with a (LC), then (RC).
   22. Press $\overline{\text{OK}}$.

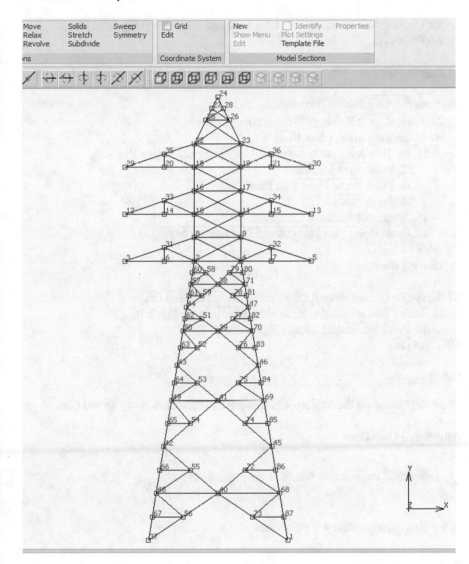

Fig. 2.21 Transmission tower mesh

Point Loads

23. Under | Boundary Conditions |: **Boundary Conditions** select New (Structural)
→ Point Load.
 24. Under Properties, tick Force Y. Set at −100.
 25. Under Entities\Nodes, press A̅d̅d̅.

26. Under ⑥ select nodes 3, 5, 12, 13, 29 and 30 with a (LC), then (RC).
27. Press OK.

Running the Job

28. Under | Jobs |: **Job** → Structural.
    29. Press Check; See in ⑧ if there are any errors.
    30. If there are none, press Run.
        31. In 'Run Job', press Advanced Job Submission (see Fig. 2.9).
            32. Press Save Model.
            33. Press Write Input File Press OK.
            34. Press Submit 1.
            35. Wait until Status = Complete.
            36. Press Open Post File (Model Plot Results Menu).

    Viewing the model

37. Under Deformed Shape\Style, select Deformed and Original.
    38. Under Scalar Plot\Style, select Numerics (see Fig. 2.10).
        39. Press Scalar and select Displacement Y.
40. Press OK.

Results

The deformation on the load application points is found to be $u_Y = -0.006$.

Additional Questions

1. Determine the reaction forces at the ground supports.

2.4   Supplementary Problems

2.4.1   *Truss Structure with Six Members*

Problem Description

Given is a plane truss structure as shown in Fig. 2.22. The members have a uniform cross-sectional area $A = 5$ and YOUNG's modulus $E = 30$. The length of each member can be taken from the figure ($L = 2$). The structure is fixed at its left-hand side and loaded by

- two prescribed displacements $u_0 = 0.2$ and $2u_0$ at the very right-hand corner,
- a vertical point load $F_0 = 30$.

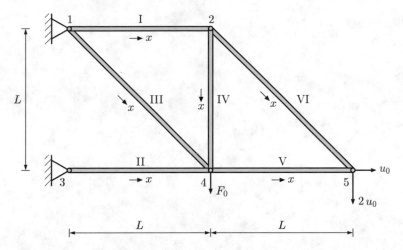

Fig. 2.22 Truss structure composed of six axial members

Model the truss structure with six linear finite elements (MSC Marc element type 9) and determine

- the displacements of the nodes,
- the reaction forces at the supports and nodes where displacements are prescribed,
- the strain, stress, and normal force in each element.
- Check the obtained results based on the global force equilibrium.

Assume that all the numbers are given in consistent units. Simplify all your results for the following special cases:

(a) $u_0 = 0$,
(b) $F_0 = 0$.

Chapter 3
Euler-Bernoulli Beams and Frames

3.1 Definition of Euler-Bernoulli Beams

The definition of EULER- BERNOULLI beam elements is summarized in Table 3.1. The derivation in lectures normally starts with the introduction of an elemental coordinate system (x, z) where the x-axis is aligned with the principal axis of the element and the z-axis is perpendicular to the element. Based on the definition of this element, displacements (u_{1z}, u_{2z}) can only occur perpendicular to the principal axis while the rotations $(\varphi_{1y}, \varphi_{2y})$ act around the y-axis.

The implementation of a beam element in a commercial finite element code is more general, i.e. based on the global coordinate system (X, Y, Z). Thus, the geometry is defined based on the global coordinates of each node (X_i, Y_i, Z_i) and the axial second moments of area I_x and I_y. The local z-axis is automatically defined from the first to the second node. This beam element is combined with a rotational element [14] and can also twist around its local z-axis. Furthermore, a rod element is superimposed. In such a configuration, each node has six degrees of freedom, i.e. the three displacements u_{iX}, u_{iY}, u_{iZ} and the three rotations $\varphi_{iX}, \varphi_{iY}, \varphi_{iZ}$ expressed in the global coordinate system. The torsional stiffness GI_p of the section is calculated as

$$ GI_p = \frac{E}{2(1 + \nu)} \left(I_x + I_y \right) , \tag{3.1} $$

which is only valid for circular cross sections[1]

---

[1]This assumption is taken if a cross sectional area A is specified in the first geometry data field. If $A = 0$ is entered in the first geometry field, Marc uses the beam section data corresponding to the section number given in the second geometry field (I_x). This allows specification of the torsional stiffness factor unequal to $I_x + I_y$ or the specification of arbitrary sections using numerical section integration, see [12].

© Springer International Publishing AG 2018
A. Öchsner and M. Öchsner, *A First Introduction to the Finite Element Analysis Program MSC Marc/Mentat*, https://doi.org/10.1007/978-3-319-71915-3_3

Table 3.1 Definition of beam elements

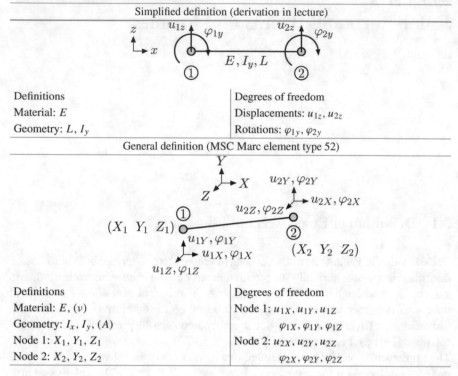

Definitions	Degrees of freedom
Material: E	Displacements: u_{1z}, u_{2z}
Geometry: L, I_y	Rotations: $\varphi_{1y}, \varphi_{2y}$

General definition (MSC Marc element type 52)

Definitions	Degrees of freedom
Material: $E, (\nu)$	Node 1: u_{1X}, u_{1Y}, u_{1Z}
Geometry: $I_x, I_y, (A)$	$\varphi_{1X}, \varphi_{1Y}, \varphi_{1Z}$
Node 1: X_1, Y_1, Z_1	Node 2: u_{2X}, u_{2Y}, u_{2Z}
Node 2: X_2, Y_2, Z_2	$\varphi_{2X}, \varphi_{2Y}, \varphi_{2Z}$

MSC Marc element type 52 uses cubic normal displacement interpolation and a linear interpolation approach for the axial displacements. In regards to strain and stress, the following outputs are obtained:

- Generalized strains:
  Axial stretch ε_z,
  Curvature about local x-axis of cross section κ_x,
  Curvature about local y-axis of cross section κ_y,
  Twist about local z-axis of cross section κ_z,

- Generalized stresses:
  Axial force,
  Bending moment about x-axis of cross section,
  Bending moment about y-axis of cross section,
  Torque about beam axis.

3.2 Basic Examples

3.2.1 Beam with a Square Cross-Section

Problem Description

Given is a beam of length $L = 10.0$ and constant bending stiffness given by $E = 200000$ and $I_y = I_z = \frac{1}{192}$[2] as shown in Fig. 3.1. At the left-hand side there is a fixed support. There are four load cases for the right-hand side of the beam:

(I) There is a point load of $F_y = -50$.
(II) There is a moment $M_z = -160$.
(III) There is a fixed displacement $u_y = -16$.
(IV) There is a rotation $\varphi_z = -1.536$.

Use a single beam element (solid section beam, type 52) to calculate the nodal unknowns.

Marc Solution

Save as 'beam_square'.

Constructing the mesh

1. Under Geometry & Mesh : **Basic Manipulation** select Geometry and Mesh (see Fig. 2.3).
   2. Under Mesh\Nodes, select $\overline{\text{Add}}$.
3. *0, 0, 0* ENTER *10, 0, 0* ENTER.
   4. Under Mesh\Elements, select $\overline{\text{Line (2)}}$ (see Fig. 2.3).
   5. $\overline{\text{Add}}$.

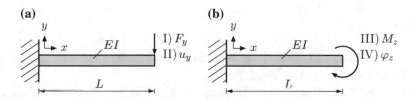

Fig. 3.1 Schematic drawing of the problem: **a** force or displacement load case; **b** moment or rotation load case

---

[2]The given second moment of area allows to calculate the side length.

Fig. 3.2 Going to the beam
geometry properties

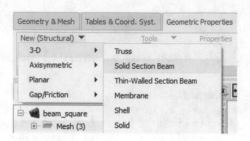

6. In ⑥, (LC) on the two new nodes.
   7. Press $\overline{OK}$.

Setting the Geometric Properties

8. Under | Geometric Properties |: **Geometric Properties** select New (Structural).
   9. Select 3D → Solid Section Beam (see Fig. 3.2).
   10. Under Properties click on Element Types. Untick Transverse Shear.
   11. Set Properties\Shape to Square and set Dimension A = 0.5 (see Fig. 3.3).
   12. Set Orientation\Vector X = 0, Y = 1, and Z = 0 (see Fig. 3.3).
   13. Under Entities\Elements, press $\overline{Add}$.
14. In ⑥ select the element with a (LC), then (RC).
   15. Press $\overline{OK}$.

Setting the Material Properties

16. Under | Material Properties |: **Material Properties** select New → Finite Stiffness
→ Standard.
   17. Set Other Properties\Young's Modulus = 200000 (see Fig. 2.6).
   18. Under Entities\Elements, press $\overline{Add}$.
19. In ⑦, press the first symbol, 'All Existing'.
   20. Press $\overline{OK}$.

Setting the Boundary Conditions

Fixed Support

21. Under | Boundary Conditions |: **Boundary Conditions** select New (Structural)
→ Fixed Displacement (see Fig. 2.7).
   22. Under Properties tick Displacement X, Y and Z, as well as Rotation X, Y
and Z.

Fig. 3.3 Setting the
geometric properties

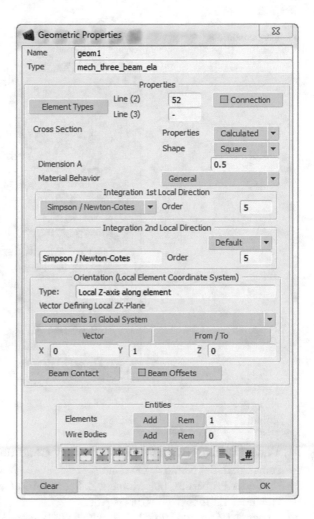

23. Under Entities\Nodes, press $\overline{Add}$.
24. Under ⑥ select the node 1 (*0, 0, 0*) with a (LC), then (RC).
   25. Press $\overline{OK}$.
   Point Load

26. Under | Boundary Conditions |: **Boundary Conditions** select New (Structural)
→ Point Load. Set Name = force1.
   27. Under Properties, tick Force Y. Set at −50.
   28. Under Entities\Nodes, press $\overline{Add}$.
29. Under ⑥ select node 2 (*10, 0, 0*) with a (LC), then (RC).
   30. Press $\overline{OK}$.

Fig. 3.4 Setting the
boundary conditions

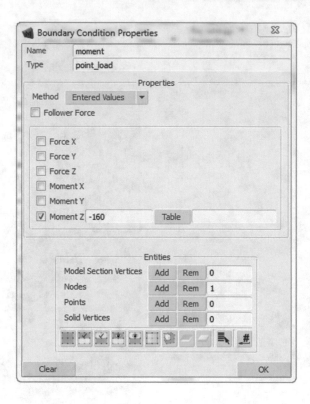

Moment

31. Under | Boundary Conditions |: **Boundary Conditions** select New (Structural)
→ Point Load. Set Name = moment.
    32. Under Properties, tick Moment Z. Set at −160 (see Fig. 3.4).
    33. Under Entities\Nodes, press $\overline{\text{Add}}$.
34. Under ⑥ select node 2 (*10, 0, 0*) with a (LC), then (RC).
    35. Press $\overline{\text{OK}}$.

Displacement

36. Under | Boundary Conditions |: **Boundary Conditions** select New (Structural)
→ Fixed Displacement. Set Name = displ1.
    37. Under Properties, tick Displacement Y. Set at −16.
    38. Under Entities\Nodes, press $\overline{\text{Add}}$.
39. Under ⑥ select node 2 (*10, 0, 0*) with a (LC), then (RC).
    40. Press $\overline{\text{OK}}$.

Rotation

41. Under ⌈Boundary Conditions⌉: **Boundary Conditions** select New (Structural)
→ Fixed Displacement. Set Name = rotation1.
    42. Under Properties, tick Rotation Z. Set at −1.536.
    43. Under Entities\Nodes, press Add.
44. Under ⑥ select node 2 (*10, 0, 0*) with a (LC), then (RC).
    45. Press OK.

Defining Loadcases

46. Under ⌈Loadcases⌉: **Loadcases** select New → Static.
    47. Set Name = force.
    48. Press Loads. Untick displ1, rotation1, moment1. Press OK.
    49. Set Stepping Procedure\#Steps = 1. Press OK.

50. Under ⌈Loadcases⌉: **Loadcases** select New → Static.
    51. Set Name = displacement.
    52. Press Loads. Untick force1, rotation1, moment1. Press OK.
    53. Set Stepping Procedure\#Steps = 1. Press OK.

54. Under ⌈Loadcases⌉: **Loadcases** select New → Static.
    55. Set Name = rotation.
    56. Press Loads. Untick displ1, force1, moment1. Press OK.
    57. Set Stepping Procedure\#Steps = 1. Press OK.

58. Under ⌈Loadcases⌉: **Loadcases** select New → Static.
    59. Set Name = moment.
    60. Press Loads. Untick displ1, force1, rotation1. Press OK.
    61. Set Stepping Procedure\#Steps = 1. Press OK.

Running the jobs

Job 1

62. Under ⌈Jobs⌉: **Job** select Structural. Set Name = fixed_force.
    63. Under Available, select force.
    64. Press Initial Loads. Untick displ1, rotation1, moment1. Press OK.
    65. Press Check; See in ⑧ if there are any errors.
    66. If there are none, press Run.
        67. In 'Run Job', press Advanced Job Submission.
            68. Press Save Model.
            69. Press Write Input File. Press OK.
            70. Press Submit 1.

71. Wait until Status = Complete. Press $\overline{OK}$
72. Press $\overline{OK}$.

Job 2

73. Under $\boxed{\text{Jobs}}$: **Job** select Structural. Set Name = fixed_displacement.
   74. Under Available, select displacement.
   75. Press $\overline{\text{Initial Loads}}$. Untick force1, rotation1, moment1. Press $\overline{OK}$.
   76. Press $\overline{\text{Check}}$; See in (8) if there are any errors.
   77. If there are none, press $\overline{\text{Run}}$.
      78. In 'Run Job', press $\overline{\text{Advanced Job Submission}}$.
         79. Press $\overline{\text{Save Model}}$.
         80. Press $\overline{\text{Write Input File}}$. Press $\overline{OK}$.
         81. Press $\overline{\text{Submit 1}}$.
      82. Wait until Status = Complete.
   83. Press $\overline{OK}$.

Job 3

84. Under $\boxed{\text{Jobs}}$: **Job** select Structural. Set Name = fixed_rotation.
   85. Under Available, select rotation.
   86. Press $\overline{\text{Initial Loads}}$. Untick displ1, force1, moment. Press $\overline{OK}$.
   87. Press $\overline{\text{Check}}$; See in ⑧ if there are any errors.
   88. If there are none, press $\overline{\text{Run}}$.
      89. In 'Run Job', press $\overline{\text{Advanced Job Submission}}$.
         90. Press $\overline{\text{Save Model}}$.
         91. Press $\overline{\text{Write Input File}}$. Press $\overline{OK}$.
         92. Press $\overline{\text{Submit 1}}$.
      93. Wait until Status = Complete.
   94. Press $\overline{OK}$.

Job 4

95. Under $\boxed{\text{Jobs}}$: **Job** select Structural. Set Name = fixed_moment.
   96. Under Available, select moment.
   97. Press $\overline{\text{Initial Loads}}$. Untick displ1, force1, rotation1. Press $\overline{OK}$.
   98. Press $\overline{\text{Check}}$; See in ⑧ if there are any errors.
   99. If there are none, press $\overline{\text{Run}}$.
      100. In 'Run Job', press $\overline{\text{Advanced Job Submission}}$.
         101. Press $\overline{\text{Save Model}}$.
         102. Press $\overline{\text{Write Input File}}$. Press $\overline{OK}$.
         103. Press $\overline{\text{Submit 1}}$.
      104. Wait until Status = Complete. Press $\overline{OK}$.
      105. Press $\overline{\text{Open Post File}}$ (Model Plot Results Menu).

Viewing the model

106. Under Deformed Shape\Style, select $\overline{\text{Deformed and Original}}$.
   107. Under Scalar Plot\Style, select $\overline{\text{Numerics}}$.
      108. Press $\overline{\text{Scalar}}$ and select Rotation Z. Press $\overline{\text{OK}}$.
   109. Press $\overline{\text{OK}}$.

Open Job beam_square_fixed_rotation.t16

110. Under File → Results → Open. Open beam_square_fixed_rotation.t16
   111. In the window 'Model Plot Results', under Scalar Plot, press $\overline{\text{Scalar}}$
      and select Reaction Moment Z.
   112. Press $\overline{\text{OK}}$.

Open Job beam_square_fixed_force.t16

113. Under File → Results → Open. Open beam_square_fixed_force.t16
   114. In the window 'Model Plot Results', under Scalar Plot, press $\overline{\text{Scalar}}$
      and select Displacement Y.
   115. Press $\overline{\text{OK}}$.

Open Job beam_square_fixed_displacement.t16

116. Under File → Results → Open. Open beam_square_fixed_displacement.t16
   117. In the window 'Model Plot Results', under Scalar Plot, press $\overline{\text{Scalar}}$ and
select Reaction Force Y.
   118. Press $\overline{\text{OK}}$.

Results

The results for the different load cases are obtained as:

(I) Point load of $F_Y = -50$: → $u_Y = -16$.
(II) Moment $M_Z = -160$: → $\varphi_Z = -1.536$.
(III) Fixed displacement $u_Y = -16$: → $F_Y = -50$.
(IV) Rotation $\varphi_Z = -1.536$: → $M_Z = -160$.

Additional Questions

1. Calculate the *analytical* solutions for the displacements $u_y(x)$ and rotations $\varphi_z(x)$
   and compare your result with the numerical solution from MSC Marc at node 2.
2. Determine in addition the *analytical* solutions for the bending moment $M_z(x)$
   and shear force $Q_y(x)$ distributions.

Fig. 3.5 Schematic drawing
of a beam with distributed
load

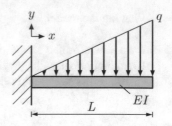

3.2.2   Beam with a Distributed Load

Problem Description

Given is a squared beam of length $L = 6.0$ and with a constant bending stiffness
given by $E = 200000$ and $I_y = I_z = \frac{1}{192}$ as shown in Fig. 3.5. At the left-hand
side there is a fixed support and on the right-hand side is free. The maximum of the
triangular shaped load is given by $q = 5$. The equivalent nodal loads for a triangled
shaped distributed load are given by:

$$F_{1y} = -\frac{3}{20}qL \,, \; F_{2y} = -\frac{7}{20}qL \,, \; M_{1y} = -\frac{qL^2}{30} \,, \; M_{2y} = \frac{qL^2}{20} \,.$$

Calculate the reaction force and moment at the left-hand and at the right-hand end
the displacement and rotation.

Marc Solution

   Save as 'beam_distr_load'.

   Constructing the mesh

1. Under Geometry & Mesh : **Basic Manipulation** select Geometry and Mesh.
   2. Under Mesh\Nodes, select $\overline{\text{Add}}$.
3. *0, 0, 0* ENTER *6, 0, 0* ENTER.
   4. Under Mesh\Elements, select $\overline{\text{Line (2)}}$.
   5. $\overline{\text{Add}}$.
6. In ⑥, (LC) on the two new nodes.
   7. Press $\overline{\text{OK}}$.

   Setting the Geometric Properties

8. Under Geometric Properties : **Geometric Properties** select New (Structural).
   9. Select 3D → Solid Section Beam (see Fig. 3.2).
   10. Under Properties click on Element Types. Untick Transverse Shear.
   11. Set Properties\Shape to Square and set Dimension A $= 0.5$.

12. Set Orientation\Vector X = 0, Y = 1, and Z = 0.
13. Under Entities\Elements, press $\overline{\text{Add}}$.
14. In ⑥ select the element with a (LC), then (RC).
   15. Press $\overline{\text{OK}}$.

Setting the Material Properties

16. Under | Material Properties |: **Material Properties** select New → Finite Stiffness
→ Standard.
   17. Set Other Properties\Young's Modulus = 200000 (see Fig. 2.6).
   18. Under Entities\Elements, press $\overline{\text{Add}}$.
19. In ⑦, press the first symbol, 'All Existing'.
   20. Press $\overline{\text{OK}}$.

Setting the Boundary Conditions

Fixed Support

21. Under | Boundary Conditions |: **Boundary Conditions** select New (Structural)
→ Fixed Displacement (see Fig. 2.7).
   22. Under Properties tick Displacement X, Y and Z, as well as Rotation
      X, Y and Z.
   23. Under Entities\Nodes, press $\overline{\text{Add}}$.
24. Under ⑥ select node 1 (0, 0, 0) with a (LC), then (RC).
   25. Press $\overline{\text{OK}}$.

Point Load 1

26. Under | Boundary Conditions |: **Boundary Conditions** select New (Structural)
→ Point Load.
   27. Under Properties, tick Force Y. Set at −10.5.
   28. Under Entities\Nodes, press $\overline{\text{Add}}$.
29. Under ⑥ select node 2 (6, 0, 0) with a (LC), then (RC).
   30. Press $\overline{\text{OK}}$.

Point Load 2 (does not affect the results, thus, could be omitted)

31. Under | Boundary Conditions |: **Boundary Conditions** select New (Structural)
→ Point Load.
   32. Under Properties, tick Force Y. Set at −4.5.
   33. Under Entities\Nodes, press $\overline{\text{Add}}$.
34. Under ⑥ select node 1 (0, 0, 0) with a (LC), then (RC).
   35. Press $\overline{\text{OK}}$.

Moment 1

36. Under ⎡Boundary Conditions⎤: **Boundary Conditions** select New (Structural)
→ Point Load.
    37. Under Properties, tick Moment Z. Set at 9.
    38. Under Entities\Nodes, press Add.
39. Under ⑥ select node 2 (*6, 0, 0*) with a (LC), then (RC).
    40. Press OK.

Moment 2 (does not affect the result! Thus, could be omitted)

41. Under ⎡Boundary Conditions⎤: **Boundary Conditions** select New (Structural)
→ Point Load.
    42. Under Properties, tick Moment Z. Set at −6.
    43. Under Entities\Nodes, press Add.
44. Under ⑥ select node 1 (*0, 0, 0*) with a (LC), then (RC).
    45. Press OK.

Running the job

46. Under ⎡Jobs⎤: **Job** select New (Structural).
    47. Press Check; See in ⑧ if there are any errors.
    48. If there are none, press Run.
        49. In 'Run Job', press Advanced Job Submission.
            50. Press Save Model.
            51. Press Write Input File. Press OK.
            52. Press Submit 1.
        53. Wait until Status = Complete. Press OK.
        54. Press Open Post File (Model Plot Results Menu).

Viewing the model

55. Under Deformed Shape\Style, select Deformed and Original.
    56. Under Scalar Plot\Style, select Numerics.
        57. Press Scalar and select Displacement Y. Press OK.
    58. Press OK.

Results

$u_{2Y} = -0.57024$.

3.2.3 *Portal Frame with a Distributed Load*

Problem Description

Given is a portal frame structure of three beam elements with length $L = 5.0$ and with a constant axial and bending stiffness given by $E = 200000$, $A = 0.5$, and $I = \frac{1}{192}$ as shown in Fig. 3.6. There is a force of magnitude $F = 100$ acting in the positive X-direction, at node 2. The equivalent nodal loads resulting from the distributed load $q_Y = -10$ are given by:

$$F_{2Y,3Y} = -\frac{qL}{2} \ , \ M_{2Y} = -\frac{qL^2}{12} \ , \ M_{3Y} = \frac{qL^2}{12} \ .$$

Calculate the reaction force and moment on the two bottom nodes and the displacement and rotation of the two top nodes.

Marc Solution

Save as 'beam_frame'.

Constructing the mesh

1. Under Geometry & Mesh : **Basic Manipulation** select Geometry and Mesh (see Fig. 2.3).
    2. Under Mesh\Nodes, select $\overline{\text{Add}}$.
3. *0, 0, 0* ENTER *0, 5, 0* ENTER *5, 5, 0* ENTER *5, 0, 0* ENTER.
    4. Under Mesh\Elements, select $\overline{\text{Line (2)}}$ (see Fig. 2.2).
    5. $\overline{\text{Add}}$.
6. In ⑥, with a (LC) connect nodes $(1, 2);(2, 3);(3, 4)$ as seen in Fig. 3.6.
    7. Press $\overline{\text{OK}}$.

Fig. 3.6 Schematic drawing of the portal frame structure

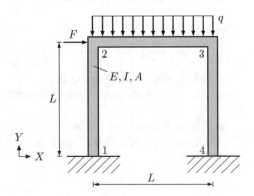

Setting the Geometric Properties—Top Beam

8. Under Geometric Properties : **Geometric Properties** select New (Structural).
   9. Select 3D → Solid Section Beam (see Fig. 3.2).
   10. Under Properties click on Element Types. Untick Transverse Shear.
   11. Set Properties\Shape to Square and set Dimension A = 0.5.
   12. Set Orientation\Vector X = 0, Y = 1, and Z = 0.
   13. Under Entities\Elements, press $\overline{\text{Add}}$.
14. In ⑥ select the top beam element with a (LC), then (RC).
   15. Press $\overline{\text{OK}}$.

Setting the Geometric Properties—Side Beams

16. Under Geometric Properties : **Geometric Properties** select New (Structural).
   17. Select 3D → Solid Section Beam (see Fig. 3.2).
   18. Under Properties click on Element Types. Untick Transverse Shear.
   19. Set Properties\Shape to Square and set Dimension A = 0.5.
   20. Set Orientation\Vector X = 0, Y = 0, and Z = 1.
   21. Under Entities\Elements, press $\overline{\text{Add}}$.
22. In ⑥ select the side beam elements with a (LC), then (RC).
   23. Press $\overline{\text{OK}}$.

Setting the Material Properties

24. Under Material Properties : **Material Properties** select New → Finite Stiffness → Standard.
   25. Set Other Properties\Young's Modulus = 200000 (see Fig. 2.6).
   26. Under Entities\Elements, press $\overline{\text{Add}}$.
27. In ⑦, press the first symbol 'All Existing'.
   28. Press $\overline{\text{OK}}$.

Setting the Boundary Conditions

Fixed Support

29. Under Boundary Conditions : **Boundary Conditions** select New (Structural) → Fixed Displacement (See Fig. 2.7).
   30. Under Properties tick Displacement X, Y and Z, as well as Rotation X, Y and Z.
   31. Under Entities\Nodes, press $\overline{\text{Add}}$.
32. Under ⑥ select node 1 (0, 0, 0) and node 4 (0, 5, 0) with a (LC), then (RC).
   33. Press $\overline{\text{OK}}$.

Point Load 1

34. Under ⟨Boundary Conditions⟩: **Boundary Conditions** select New (Structural)
→ Point Load.
    35. Under Properties, tick Force X. Set at 100.
    36. Under Entities\Nodes, press Add.
37. Under ⑥ select node 2 (*0, 5, 0*) with a (LC), then (RC).
    38. Press OK.

Point Load 2

39. Under ⟨Boundary Conditions⟩: **Boundary Conditions** select New (Structural)
→ Point Load.
    40. Under Properties, tick Force Y. Set at -25.
    41. Under Entities\Nodes, press Add.
42. Under ⑥ select node 2 (*0, 5, 0*) and node 3 (*5, 5, 0*) with a (LC), then (RC).
    43. Press OK.

Moment 1

44. Under ⟨Boundary Conditions⟩: **Boundary Conditions** select New (Structural)
→ Point Load.
    45. Under Properties, tick Moment Z. Set at -20.8333.
    46. Under Entities\Nodes, press Add.
47. Under ⑥ select node 2 (*0, 5, 0*) with a (LC), then (RC).
    48. Press OK.

Moment 2

49. Under ⟨Boundary Conditions⟩: **Boundary Conditions** select New (Structural)
→ Point Load.
    50. Under Properties, tick Moment Z. Set at 20.8333.
    51. Under Entities\Nodes, press Add.
52. Under ⑥ select node 3 (*5, 5, 0*) with a (LC), then (RC).
    53. Press OK.

Running the job

54. Under ⟨Jobs⟩: **Job** select New (Structural).
    55. Press Check; See in ⑧ if there are any errors.
    56. If there are none, press Run.
        57. In 'Run Job', press Advanced Job Submission.
        58. Press Save Model.
        59. Press Write Input File. Press OK.

60. Press $\overline{\text{Submit 1}}$.
61. Wait until Status = Complete. Press $\overline{\text{OK}}$.
62. Press $\overline{\text{Open Post File}}$ (Model Plot Results Menu).

Viewing the model

63. Under Deformed Shape\Style, select $\overline{\text{Deformed and Original}}$.
64. Under Scalar Plot\Style, select $\overline{\text{Numerics}}$.
65. Press $\overline{\text{Scalar}}$ and select Displacement Y. Press $\overline{\text{OK}}$.
66. Press $\overline{\text{Scalar}}$ and select Rotation Z. Press $\overline{\text{OK}}$.
67. Press $\overline{\text{OK}}$.

Results

The results for the displacements and the reactions are obtained as:

$$u_{2X} = 0.72065\,,\ u_{2Y} = 0.0017735\,,\ \varphi_{2Z} = -0.104386\,,$$
$$u_{3X} = 0.715247\,,\ u_{3Y} = -0.0067735\,,\ \varphi_{3Z} = -0.0699728\,,$$
$$F_{R1X} = -45.9685\,,\ F_{R1Y} = -17.735\,,\ M_{R1Z} = 136.668\,,$$
$$F_{R4X} = -54.0315\,,\ F_{R4Y} = 67.735\,,\ M_{R4Z} = 149.656\,.$$

3.3   Advanced Examples

3.3.1   *Plane Bridge Structure with Beam Elements*

Problem Description

Recalculate problem '2.3.1 Plane Bridge Structure' based on beam elements (i.e., thin elastic beams, MSC Marc element type 52) maintaining the geometry. Assume a squared cross-sectional area of 10. Modify the existing model from problem 2.3.1 to calculate the deformations at node 3 for the vertical force case $F_Y = -100$ at node 3. Compare your results to the values obtained from the truss structure and comment on the differences.

Consider the following steps to modify the model from problem 2.3.1:

1. Under $\boxed{\text{Geometric Properties}}$: **Geometric Properties** select Properties.
2. Under Entities\Elements, press $\overline{\text{Rem}}$.
3. In ⑦, press the first symbol 'All Existing'.
4. Press $\overline{\text{OK}}$.

5. Under | Geometric Properties |: **Geometric Properties** select Tools and Remove Unused.
6. Under | Geometric Properties |: **Geometric Properties** select New (Structural).
 7. Select 3D → Solid Section Beam (see Fig. 3.2).
 8. Under Properties click on Element Types. Untick Transverse Shear.
 9. Set Properties\Shape to Square and set Dimension A = 3.162277660.
 10. Set Orientation\Vector X = 0, Y = 0, and Z = 1.
 11. Under Entities\Elements, press $\overline{\text{Add}}$.
12. In ⑦, press the first symbol 'All Existing'.
 13. Press $\overline{\text{OK}}$.
14. Under | Boundary Conditions |: **Boundary Conditions** select Edit, press $\overline{\text{apply 1}}$ and $\overline{\text{OK}}$.
15. Under | Boundary Conditions |: **Boundary Conditions** select Properties
 16. Under Properties tick Displacement Z.
 17. Press $\overline{\text{OK}}$.
18. Under | Jobs |: **Element Types** select Element Types.
 19. Select Truss/Beam → press $\overline{52}$
20. In ⑦, press the first symbol 'All Existing'.
 21. Press $\overline{\text{OK}}$. Press $\overline{\text{OK}}$.

Solve the problem in the common way to answer the question.

3.3.2 Transmission Tower with Beam Elements

Problem Description

Recalculate problem '2.3.2 Transmission Tower Structure' based on beam elements (i.e., thin elastic beams, MSC Marc element type 52) under maintaining the geometry. Assume a squared cross-sectional area of 0.5. Modify the existing model from problem 2.3.2 to calculate the vertical deformations at the load application points and the reaction forces at the supports. Compare your results to the values obtained from the truss structure and comment on the differences.

Consider the following steps to modify the model from problem 2.3.2:

1. Under | Geometric Properties |: **Geometric Properties** select Properties.
 2. Under Entities\Elements, press $\overline{\text{Rem}}$.
3. In ⑦, press the first symbol 'All Existing'.
 4. Press $\overline{\text{OK}}$.
5. Under | Geometric Properties |: **Geometric Properties** select Tools and Remove Unused.

6. Under $\boxed{\text{Geometric Properties}}$: **Geometric Properties** select New (Structural).
    7. Select 3D → Solid Section Beam (see Fig. 3.2).
    8. Under Properties click on Element Types. Untick Transverse Shear.
    9. Set Properties\Shape to Square and set Dimension A = 3.162277660.
    10. Set Orientation\Vector X = 0, Y = 0, and Z = 1.
    11. Under Entities\Elements, press $\overline{\text{Add}}$.
12. In ⑦, press the first symbol 'All Existing'.
    13. Press $\overline{\text{OK}}$.
14. Under $\boxed{\text{Boundary Conditions}}$: **Boundary Conditions** select Edit, press $\overline{\text{apply 1}}$
and $\overline{\text{OK}}$.
15. Under $\boxed{\text{Boundary Conditions}}$: **Boundary Conditions** select Properties
    16. Under Properties tick Displacement Z.
    17. Press $\overline{\text{OK}}$.
18. Under $\boxed{\text{Jobs}}$: **Element Types** select Element Types.
    19. Select Truss/Beam → press $\overline{52}$
20. In ⑦, press the first symbol 'All Existing'.
    21. Press $\overline{\text{OK}}$. Press $\overline{\text{OK}}$.
Solve the problem in the common way to answer the question.

3.3.3   Beam Element—Generalized Strain and Stress Output

Problem Description

Given is a cantilever beam of length $L = 10.0$ and constant bending stiffness given
by $E = 200000$ and $I_y = I_z = \frac{1}{192}$ (assume a square cross section with side length
a) as shown in Fig. 3.7. At the left-hand side there is a fixed support. There are three
point loads acting at the right-hand side of the beam:

(I) A point load in positive x-direction: $F_x = 10$,
(II) a point load in negative y-direction: $F_y = -50$,
(III) a point load in negative z-direction: $F_z = -25$.

Fig. 3.7 Cantilever beam
loaded by different point
loads

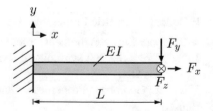

• Consider first a simplified case where *only* the vertical force F_y is acting and calculate the *analytical* solution for:

– The bending line $u_y(x)$,
– the bending moment distribution $M_z(x)$,
– the shear force distribution $Q_y(x)$,
– the curvature $\kappa_z(x)$,
– the maximum normal stress at both ends $|\sigma_{max}|$,
– the maximum normal strain at both ends $|\varepsilon_{max}|$.

Derive first the general analytical solution based on E, I_z, L, a, F_y and then simplify based on the given numerical values.

• Calculate the numerical solution of the problem based on a single beam element (i.e., thin elastic beam, MSC Marc element type 52) and evaluate all the strains and stresses at the nodes. Conclude based on the analytical calculation which quantities are shown for the stresses and strains. Compare the 'stress' and 'strain' values from MSC Marc to the analytical values. Why is there a small difference?

3.3.4 Beam Element—Generalized Stresses for Multiaxial Stress State

Problem Description

Given is a cantilever beam of length $L = 10.0$ and constant bending stiffness given by $E = 200000$ ($\nu = 0.3$) and $I_y = I_z = \frac{1}{192}$ (assume a circular cross section with diameter a) as shown in Fig. 3.8. At the left-hand side there is a fixed support. There are four different point loads acting at the right-hand side of the beam:

(a) A point load in negative y-direction: $F_y = -50$,
(b) a point load in negative z-direction: $F_z = -25$,
(c) a point load in positive x-direction: $F_x = 10$,
(d) a single moment around the positive x-axis: $M_x = 125$.

• Consider first each load case separately and calculate the analytical solution for the stresses. Identify for each load case the location of the maximum stress. Consider in a second step that all load cases are simultaneously acting. Identify the location of the maximum stress according to the VON MISES yield condition.

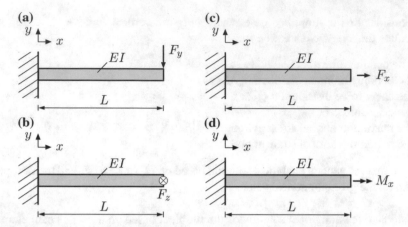

Fig. 3.8 Cantilever beam loaded by different point loads: **a** bending in x-y plane, **b** bending in x-z plane, **c** tension and **d** torsion

• Calculate the numerical solution of the problem based on two beam elements of equal length (i.e., thin elastic beam, MSC Marc element type 52) and evaluate the (generalized) stresses and the equivalent value. Compare and relate both approaches.

3.3.5 Chassis of a Formula SAE Racing Car

Problem Description

Many universities take part in the Formula SAE[3] design competition. Formula SAE is a student challenge to conceive, design, fabricate, and compete with small formula-style racing cars [24]. A typical chassis of such a car is shown in Fig. 3.9.

The following steps are a few hints on how such a structure could be modeled based on one-dimensional beam elements. Complex geometries are normally created in an external CAD package and the corresponding geometry file can be imported into Marc/Mentat. Let us assume that the geometry is available in the neutral file format IGES. Then, the procedure

File → Import → IGES...

allows to import the file which contains the geometry (see Fig. 3.10). It may happen during the import of the file that additional points are created.

---

[3]SAE stands for Society of Automotive Engineers.

Fig. 3.9 Schematic
representation of a chassis

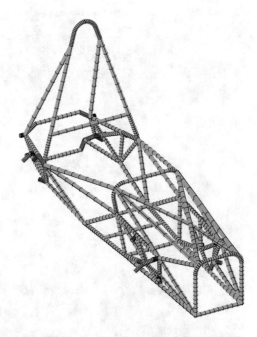

Fig. 3.10 Geometry import
menu for IGES file format

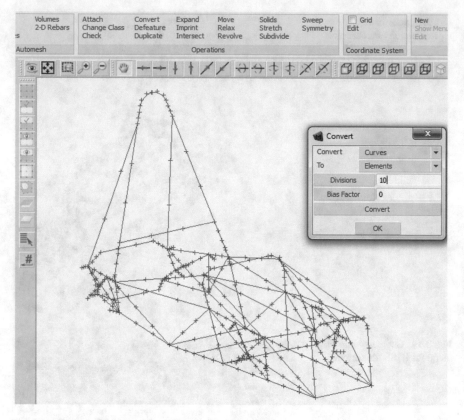

Fig. 3.11 Imported geometry based on the IGES file

The more challenging step is the conversion of the pure geometry into a finite element mesh. Let us assume that the geometry is solely represented by points and curves[4] (see Fig. 3.11) and that we intend to model the problem based on one-dimensional beam elements.
The steps

> Geometry & Mesh : **Operations** select Convert (see Fig. 3.11)

allow to convert the geometry (in this case 'Curves') into elements. The division factor defines how many elements are used for a single curve. This allows, together

---

[4]The user must pay attention to the fact that not all curves might be properly connected and these gaps must be fixed.

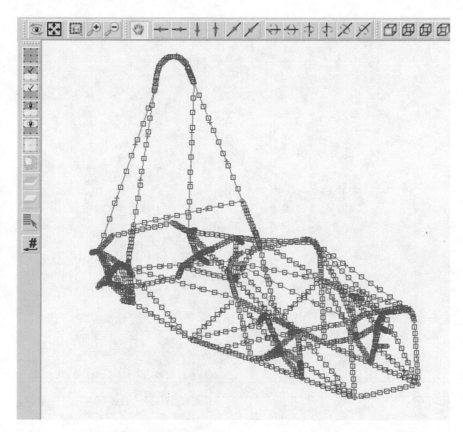

Fig. 3.12 Imported IGES geometry converted into a mesh based on one-dimensional elements (beams)

with the 'Bias Factor', to approximate the geometry with (straight!) elements (see Fig. 3.12). Finally, it might be required to use the 'Sweep' command to properly connect the single elements.

Chapter 4
Timoshenko Beams

4.1 Definition of Timoshenko Beams

The definition of TIMOSHENKO beam elements, which allows for the consideration of transverse shear effects, is summarized in Table 4.1. The derivation in lectures normally starts with the introduction of an elemental coordinate system (x, z) where the x-axis is aligned with the principal axis of the element and the z-axis is perpendicular to the element. Based on the definition of this element, displacements (u_{1z}, u_{2z}) can only occur perpendicular to the principal axis while the rotations (ϕ_{1y}, ϕ_{2y}) act around the y-axis.

The implementation of a beam element in a commercial finite element code is more general, i.e. based on the global coordinate system (X, Y, Z). Thus, the geometry is defined based on the global coordinates of each node (X_i, Y_i, Z_i), the axial second moments of area I_x and I_y, and the shear areas $A_{s,x}$ and $A_{s,y}$. The local z-axis is automatically defined from the first to the second node. This beam element is combined with a rotational element [14] and can also twist around its local z-axis. Furthermore, a rod element is superimposed. In such a configuration, each node has six degrees of freedom, i.e. the three displacements u_{iX}, u_{iY}, u_{iZ} and the three rotations $\phi_{iX}, \phi_{iY}, \phi_{iZ}$ expressed in the global coordinate system. The torsional stiffness GI_p of the section is calculated as

$$GI_p = \frac{E}{2(1 + \nu)} \left(I_x + I_y \right) , \tag{4.1}$$

which is only valid for circular cross sections.[1]

---

[1] This assumption is taken if a cross sectional area A is specified in the first geometry data field. If $A = 0$ is entered in the first geometry field, Marc uses the beam section data corresponding to the section number given in the second geometry field (I_x). This allows specification of the torsional stiffness factor unequal to $I_x + I_y$ as well as the shear areas $A_{s,x}$ and $A_{s,y}$ unequal to the area A or the specification of arbitrary sections using numerical section integration, see [12].

© Springer International Publishing AG 2018   65
A. Öchsner and M. Öchsner, *A First Introduction to the Finite Element Analysis Program MSC Marc/Mentat*, https://doi.org/10.1007/978-3-319-71915-3_4

Table 4.1 Definition of TIMOSHENKO beam elements

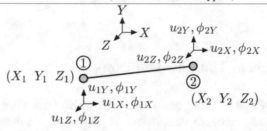

Simplified definition (derivation in lecture)	
Definitions	Degrees of freedom
Material: E, G	Displacements: u_{1z}, u_{2z}
Geometry: L, A, I_y, k_s	Rotations: ϕ_{1y}, ϕ_{2y}

General definition (MSC Marc element type 9)

Definitions	Degrees of freedom
Material: E, ν	Node 1: u_{1X}, u_{1Y}, u_{1Z}
	$\phi_{1X}, \phi_{1Y}, \phi_{1Z}$
Geometry: I_x, I_y, A $A_{s,x}, A_{s,y}$	Node 2: u_{2X}, u_{2Y}, u_{2Z}
	$\phi_{2X}, \phi_{2Y}, \phi_{2Z}$
Node 1: X_1, Y_1, Z_1	
Node 2: X_2, Y_2, Z_2	

MSC Marc element type 98 uses linear interpolation[2] for the transverse and axial displacements as well as for the rotations. In regards to strain and stress, the following outputs are obtained:

- Generalized strains:
  Axial stretch ε_z,
  Local γ_{xz} shear strain,
  Local γ_{yz} shear strain,
  Curvature about local x-axis of cross section κ_x,
  Curvature about local y-axis of cross section κ_y,
  Twist about local z-axis of cross section κ_z,

---

[2]This element uses a one-point integration rule for the stiffness matrix.

- Generalized stresses:
  Axial force,
  Shear force in local x-direction,
  Shear force in local y-direction,
  Bending moment about x-axis of cross section,
  Bending moment about y-axis of cross section,
  Torque about beam axis.

4.2 Basic Example

4.2.1 Timoshenko Beam with a Square Cross-Section

Problem Description

A cantilever beam is loaded by a single force $F = 100$ in the negative y-direction at its right-hand end (see Fig. 4.1). The height of the beam's square cross-section is given by $h = 0.5$ and the beam has constant material properties given by $E = 200000$ and a Poisson's ratio of 0.3. There are two cases:
(I) The length L of the beam is $L = 10 \times h$.
(II) The length L of the beam is $L = \frac{h}{5}$.

For each case calculate the maximum deformation of the right-hand end using the following number of TIMOSHENKO beams:

(a) 1 element,
(b) 5 elements,
(c) 10 elements,
(d) 50 elements.

Compare your results with the *analytical* solution for the EULER- BERNOULLI and TIMOSHENKO beam. In addition, compare your results with a finite element approach based on EULER- BERNOULLI beam elements.

Fig. 4.1 Schematic drawing of the problem

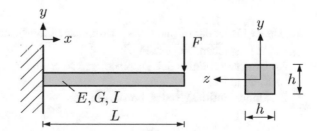

Marc Solution

Save as 'timo_long_1'

Constructing the mesh

1. Under Geometry & Mesh : **Basic Manipulation** select Geometry and Mesh (see Fig. 2.2).
    2. Under Mesh\Nodes, select $\overline{\text{Add}}$.
3. *0,0,0* ENTER *5,0,0* ENTER.
    4. Under Mesh\Elements, select $\overline{\text{Line (2)}}$ (see Fig. 2.2).
    5. $\overline{\text{Add}}$.
6. In ⑥, (LC) on the two new nodes.
    7. Press $\overline{\text{OK}}$.
Setting the Geometric Properties

8. Under Geometric Properties : **Geometric Properties** select New (Structural).
    9. Select 3D → Solid Section Beam (see Fig. 3.2).
    10. Set Properties\Shape to Square and set Dimension A = 0.5.
    11. Set Orientation\Vector X = 0, Y = 0 and Z = 1.
    12. Under Entities\Elements, press $\overline{\text{Add}}$.
13. In ⑥ select the element with a (LC), then (RC).
    14. Press $\overline{\text{OK}}$.

Setting the Material Properties

15. Under Material Properties : **Material Properties** select New → Finite Stiffness Region → Standard.
    16. Set Other Properties\Young's modulus = 200000 (see Fig. 2.6).
    17. Set Other Properties\Poisson's ratio = 0.3.
    18. Under Entities\Elements, press $\overline{\text{Add}}$.
19. In ⑥ select the element with a (LC), then (RC).
    20. Press $\overline{\text{OK}}$.

Setting the Boundary Conditions

Fixed Support

21. Under Boundary Conditions : **Boundary Conditions** select New (Structural) → Fixed Displacement (See Fig. 2.7).
    22. Under Properties tick Displacement X, Y and Z, as well as Rotation X, Y and Z.
    23. Under Entities\Nodes, press $\overline{\text{Add}}$.

24. Under ⑥ select node 1 (*0,0,0*) with a (LC), then (RC).
25. Press $\overline{OK}$.

Point Load

26. Under | Boundary Conditions |: **Boundary Conditions** select New (Structural) → Point Load.
  27. Under Properties, tick Force Y. Set at −100.
  28. Under Entities\Nodes, press $\overline{Add}$.
29. Under ⑥ select node 2 (*5,0,0*) with a (LC), then (RC).
  30. Press $\overline{OK}$.

Running the Jobs

31. Under | Jobs |: **Job** select New → Structural.
  32. Press $\overline{Check}$; See in ⑧ if there are any errors.
  33. If there are none, press $\overline{Run}$.
    34. In 'Run Job', press $\overline{Advanced\ Job\ Submission}$.
      35. Press $\overline{Save\ Model}$.
      36. Press $\overline{Write\ Input\ File}$. Press $\overline{OK}$.
      37. Press $\overline{Submit\ 1}$.
      38. Wait until Status = Complete.
      39. Press $\overline{Open\ Post\ File}$ (Model Plot Results Menu).

Viewing the model

40. Under Deformed Shape\Style, select $\overline{Deformed\ and\ Original}$.
  41. Under Scalar Plot\Style, select $\overline{Numerics}$.
    42. Press $\overline{Scalar}$ and select Displacement Y.
  43. Press $\overline{OK}$.

For cases I.b-c and II.b-d, add these steps after step 7, then continue with step 8 above as with case I.a:

8. Under | Geometry & Mesh |: **Operations** select Subdivide, see Fig. 4.2.
  9. Set the first three rows to 5; 1; 1, for case b), 10; 1; 1, for case c), or 50; 1; 1, for case d).
  10. Press $\overline{Elements}$, then in ⑥ select the element with a (LC), then (RC). Press $\overline{OK}$.
  11. Under | Geometry & Mesh |: **Operations** select Sweep.
  12. Set Sweep/Tolerance to 0.01 (see Fig. 4.3).
  13. Under Sweep, press $\overline{All}$. Press $\overline{OK}$.

Results
Collect the results for this problem in Table 4.2.

Fig. 4.2 Using the
subdivide function

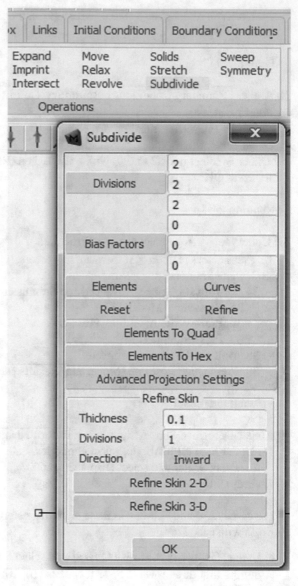

4.3   Advanced Example

4.3.1   Beam Element—Generalized Strain and Stress Output

Problem Description

A short cantilever beam is loaded by a single force $F = 100$ in the positive y-direction
at its right-hand end, see Fig. 4.4. The height of the beam's square cross-section is

Fig. 4.3 Sweeping to remove duplicate nodes

Table 4.2 Comparison of numerical and analytical results

	Case I, $L = 10 \times h$		Case II, $L = 0.2 \times h$	
	Euler-Bernoulli	Timoshenko	Euler-Bernoulli	Timoshenko
Analytical solution	-4	x.xxxx	x.xxxxxx	x.xxxxxx
u_Y: 1 element	x	-3.026	x.xxxxxx	x.xxxxxx
rel. error[a] in %	x%	xx.xxx%	xx.xxx%	xx.xxxx%
u_Y: 5 elements	x	x.xxx	x.xxxxxx	x.xxxxxx
rel. error in %	x%	x.xxx%	xx.xxx%	xx.xxx%
u_Y: 10 elements	x	x.xxx	x.xxxxxx	x.xxxxxx
rel. error in %	x%	x.xxx%	xx.xxx%	xx.xxx%
u_Y: 50 elements	x	x.xxx	x.xxxxxx	x.xxxxxxxxxx
rel. error in %	x%	x.xxx%	xx.xxx%	xx.xxx%

[a]The relative error is calculated based on the formula $|\frac{\text{analytical solution - FE solution}}{\text{analytical solution}}| \times 100$. For the $L = 0.2 \times h$ case, the Timoshenko solution was taken as the analytical solution for both cases.

Fig. 4.4 Schematic drawing
of the short beam problem

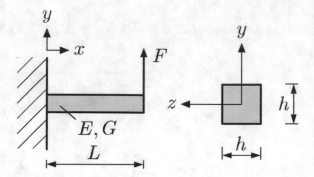

given by $h = 0.5$ and the beam has constant material properties given by $E = 200000$
and a Poisson's ratio of 0.3. The length of the beam equals $L = 0.2h$.
Calculate the problem based on a single beam element (i.e., thick elastic beam, MSC
Marc element type 98) and evaluate all the strains and stresses at the nodes. Conclude
based on the analytical calculation which quantities are shown for the stresses and
strains.

Marc Solution

Save as 'timo_short' and follow the steps of example in Sect. 4.2.1. However,
modify the step 'Setting the Geometric Properties' as follows (see Fig. 4.5):

8. Under | Geometric Properties |: **Geometric Properties** select New (Structural).
   9. Select 3D → Solid Section Beam (see Fig. 3.2).
   10. Set Properties to Entered and set Area = 0.25, Ixx = 0.00520833, Iyy =
0.00520833, see Fig. 4.5.
   11. Under Properties tick Additional Cross Section Properties and set Effective
Transverse Shear Area Ax = 0.208333 and Effective Transverse Shear Area Ay =
0.208333.
   12. Set Orientation\Vector X = 0, Y = 0 and Z = 1.
...

Comment: The geometrical properties are calculated as follows:

- Cross-sectional area: $A = h^2$,
- second moment of area: $I_{xx} = \frac{h^4}{12}$,
- second moment of area: $I_{yy} = \frac{h^4}{12}$,
- transverse shear area: $A_x = k_s A$ (shear correction factor for squared cross section:
  $k_s = \frac{5}{6}$),
- transverse shear area: $A_y = k_s A$.

Fig. 4.5 Setting the
geometric properties

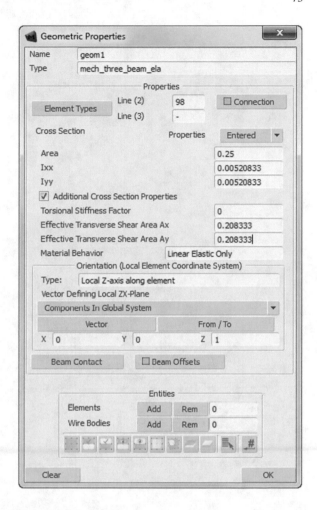

In addition, change the 'Running the Jobs' procedure to request the stress and strain
output:

31. Under ⟨ Jobs ⟩ : **Job** select New → Structural.
    32. Press Job Results. Under Available Element Tensors, tick Stress and
        Total Strain, and then press OK.

Chapter 5
Plane Elements

5.1 Definition of Plane Elements

The definition of a *plane stress* element is summarized in Table 5.1. This four-node, isoparametric quadrilateral uses bilinear interpolation functions and the stiffness matrix is formed using four-point Gaussian integration. The node numbering must follow the scheme shown in Table 5.1.

The strains $(\varepsilon_X, \varepsilon_Y, \gamma_{XY})$ and the stresses $(\sigma_X, \sigma_Y, \sigma_{XY})$ are evaluated at the four integration points (or the centroid of the element). The thickness strain ε_Z is not printed and shown as 0 for isotropic materials but can be simply calculated as

$$\varepsilon_Z = -\tfrac{\nu}{E}(\sigma_X + \sigma_Y). \tag{5.1}$$

The definition of a *plane strain* element (MSC Marc element type 11) is similar to the definition for the plane stress element as summarized in Table 5.1. The strains $(\varepsilon_X, \varepsilon_Y, \gamma_{XY})$ and the stresses $(\sigma_X, \sigma_Y, \sigma_Z, \sigma_{XY})$ are evaluated at the four integration points (or the centroid of the element).

5.2 Basic Examples

5.2.1 Plane Element Under Tensile Load

Problem Description

Given is a regular two-dimensional plane stress element as shown in Fig. 5.1, with a width of $2a$, a height of $2b$ and a thickness of h. The left-hand nodes are fixed and the right-hand nodes are loaded by a horizontal point load $F = 100$. Use a single plane elasticity element to calculate the nodal displacements for $a = 0.75$, $b = 0.5$, $t = 0.05$, $E = 200000$ and $\nu = 0.2$.

© Springer International Publishing AG 2018
A. Öchsner and M. Öchsner, *A First Introduction to the Finite Element Analysis Program MSC Marc/Mentat*, https://doi.org/10.1007/978-3-319-71915-3_5

Table 5.1 Definition of a plane stress quadrilateral element

General Definition (MSC Marc element type 3)

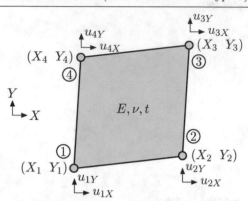

Definitions	Degrees of Freedom
Material: E, ν	Node 1: u_{1X}, u_{1Y}
Geometry: t	Node 2: u_{2X}, u_{2Y}
Node 1: X_1, Y_1	Node 3: u_{3X}, u_{3Y}
Node 2: X_2, Y_2	Node 4: u_{4X}, u_{4Y}
Node 3: X_3, Y_3	
Node 4: X_4, Y_4	

Fig. 5.1 Schematic drawing
of the problem

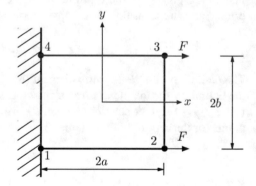

Marc Solution

Save as 'plate_tensile'.

Constructing the mesh

1. Under Geometry & Mesh : **Basic Manipulation** select Geometry and Mesh (see Fig. 2.3).

2. Under Mesh\Nodes, select $\overline{\text{Add}}$.

3. *0,0,0* ENTER *1.5,0,0* ENTER *1.5,1,0* ENTER *0,1,0* ENTER.

4. Under Mesh\Elements, select $\overline{\text{Quad (4)}}$ (see Fig. 5.2).

5. $\overline{\text{Add}}$.

Fig. 5.2 Setting the element type

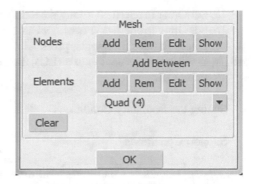

6. In ⑥, click on nodes 1–4, beginning with the bottom left node (1) and continuing the selection in counter-clockwise direction.
7. Press $\overline{\text{OK}}$.

Setting the Geometric Properties

8. Under Geometric Properties : **Geometric Properties** select New (Structural).
    9. Select → Planar → Plane Stress (see Fig. 5.3).
    10. Set Properties\Thickness = 0.05.
    11. Under Entities\Elements, press $\overline{\text{Add}}$.
12. In ⑥ select the element with a (LC), then (RC).
    13. Press $\overline{\text{OK}}$.

Setting the Material Properties

14. Under Material Properties : **Material Properties** select New → Finite Stiffness Region → Standard.

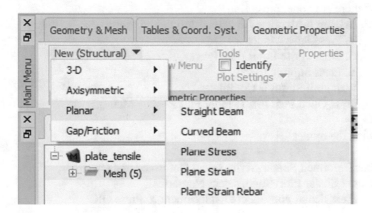

Fig. 5.3 Going to the plane geometry properties

15. Set Other Properties\Young's Modulus = 200000 (see Fig. 2.6).
16. Set Other Properties\Poisson's Ratio = 0.2.
17. Under Entities\Elements, press $\overline{\text{Add}}$.
18. In ⑥ select the element with a (LC), then (RC).
    19. Press $\overline{\text{OK}}$.

Setting the Boundary Conditions

Support Conditions

20. Under $\boxed{\text{Boundary Conditions}}$: **Boundary Conditions** select New (Structural)
→ Fixed Displacement (see Fig. 2.7).
    21. Under Properties tick Displacement X, Y and Z.
    22. Under Entities\Nodes, press $\overline{\text{Add}}$.
23. Under ⑥ select node 1 *(0,0,0)* and node 4 *(0,1,0)* with a (LC), then (RC).
    24. Press $\overline{\text{OK}}$.

Point Load

25. Under $\boxed{\text{Boundary Conditions}}$: **Boundary Conditions** select New (Structural)
→ Point Load.
    26. Under Properties, tick Force X. Set at 100.
    27. Under Entities\Nodes, press $\overline{\text{Add}}$.
28. Under ⑥ select node 2 *(1.5,0,0)* and node 3 *(1.5,1,0)* with a (LC), then (RC).
    29. Press $\overline{\text{OK}}$.

Running the Jobs

30. Under $\boxed{\text{Jobs}}$: **Job** select Structural.
    31. Press $\overline{\text{Check}}$; See in ⑧ if there are any errors.
    32. If there are none, press $\overline{\text{Run}}$.
        33. In 'Run Job', press $\overline{\text{Advanced Job Submission}}$.
            34. Press $\overline{\text{Save Model}}$.
            35. Press $\overline{\text{Write Input File}}$. Press $\overline{\text{OK}}$.
            36. Press $\overline{\text{Submit 1}}$.
        37. Wait until Status = Complete.
        38. Press $\overline{\text{Open Post File}}$ (Model Plot Results Menu).

Viewing the model

39. Under Deformed Shape\Style, select $\overline{\text{Deformed and Original}}$.
    40. Under Scalar Plot\Style, select $\overline{\text{Numerics}}$.
        41. Press $\overline{\text{Scalar}}$ and select Displacement Y. Press $\overline{\text{OK}}$.

Table 5.2 Summary of numerical results for the *plane stress* problem

Quantity	Node 2	Node 3
u_X	0.0296517	0.0296517
u_Y	X.XXXXXX	X.XXXXXX
σ_X	X.XXXXXX	X.XXXXXX
σ_Y	X.XXXXXX	X.XXXXXX
ε_Z	X.XXXXXX	X.XXXXXX
σ_Z	X.XXXXXX	X.XXXXXX

42. Press $\overline{\text{Scalar}}$ and select Displacement X. Press $\overline{\text{OK}}$.
43. Press $\overline{\text{OK}}$.

Results

Collect the results for this problem in Table 5.2.
Repeat the same problem for a *plane strain* element and collect the same data as indicated in Table 5.2.

5.2.2 Simply Supported Beam

Problem Description

Given is a simply supported beam as indicated in Fig. 5.4. The length of the beam is $4a$ and the rectangular cross section has the dimensions $h \times 2b$. The beam is loaded by a single force $F = 100$ acting in the middle of the beam. Note that the problem is not symmetric. Use plane elasticity two-dimensional elements in the following to model the problem and to calculate the nodal unknowns under the assumption of a plane *stress* case. The modelling approach is based on two elements with nodes 1, ..., 6, see Fig. 5.5. Consider $a = \frac{3}{4}$, $b = \frac{1}{2}$, $t = 0.05$ and $E = 200000$, $\nu = 0.2$.

a) Calculate the numerical solution for all nodal displacements at $x = 2a$ under the assumption that the force F is acting at node 3.

Fig. 5.4 Schematic drawing of the problem

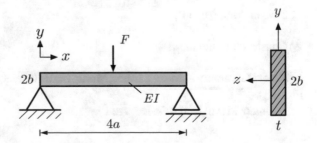

Fig. 5.5 Schematic drawing
of the different modelling
approaches for a simply
supported beam

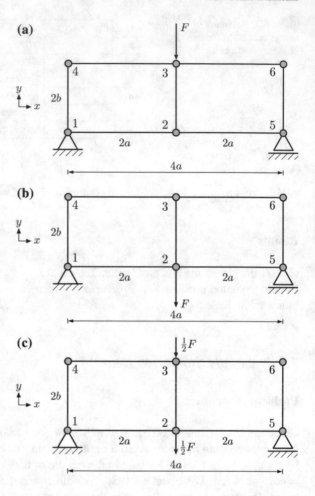

b) Calculate the numerical solution for all nodal displacements at $x = 2a$ under the
assumption that the force F is acting at node 2.

c) Calculate the numerical solution for all nodal displacements at $x = 2a$ under the
assumption that the force $\frac{1}{2}F$ is acting at node 3 and $\frac{1}{2}F$ is acting at node 2.

Marc Solution

Save as 'plate_bend'.

Constructing the mesh

1. Under Geometry & Mesh : **Basic Manipulation** select Geometry and Mesh (see
Fig. 2.3).
    2. Under Mesh\Nodes, select $\overline{\text{Add}}$.

3. Nodes: 1.*(0,0,0)* 2.*(1.5,0,0)* 3.*(1.5,1,0)* 4.*(0,1,0)* 5.*(3,0,0)* 6.*(3,1,0)*.
   4. Under Mesh\Elements, select $\overline{\text{Quad (4)}}$ (see Fig. 5.2).
   5. $\overline{\text{Add}}$.
6. In ⑥, click on nodes 1-4, beginning with the bottom left node (1) and continuing the selection in counter-clockwise direction. Then select nodes 2, 5, 6, 3.
   7. Press $\overline{\text{OK}}$.

Setting the Geometric Properties

8. Under | Geometric Properties |: **Geometric Properties** select New (Structural).
   9. Select → Planar → Plane Stress (see Fig. 5.3).
   10. Set Properties\Thickness = 0.05.
   11. Under Entities\Elements, press $\overline{\text{Add}}$.
12. In ⑥ select both elements with a (LC), then (RC).
   13. Press $\overline{\text{OK}}$.

Setting the Material Properties

14. Under | Material Properties |: **Material Properties** select New → Finite Stiffness Region → Standard.
   15. Set Other Properties\Young's Modulus = 200000 (see Fig. 2.6).
   16. Set Other Properties\Poisson's Ratio = 0.2.
   17. Under Entities\Elements, press $\overline{\text{Add}}$.
18. In ⑥ select both elements with a (LC), then (RC).
   19. Press $\overline{\text{OK}}$.

Setting the Boundary Conditions

Support Conditions

20. Under | Boundary Conditions |: **Boundary Conditions** select New (Structural) → Fixed Displacement (see Fig. 2.7).
   21. Under Properties tick Displacement X, Y and Z.
   22. Under Entities\Nodes, press $\overline{\text{Add}}$.
23. Under ⑥ select node 1 *(0,0,0)* with a (LC), then (RC).
   24. Press $\overline{\text{OK}}$.

25. Under | Boundary Conditions |: **Boundary Conditions** select New (Structural) → Fixed Displacement (see Fig. 2.7).
   26. Under Properties tick Displacement Y and Z.
   27. Under Entities\Nodes, press $\overline{\text{Add}}$.
28. Under ⑥ select node 5 *(3,0,0)* with a (LC), then (RC).
   29. Press $\overline{\text{OK}}$.

Point Loads

30. Under Boundary Conditions : **Boundary Conditions** select New (Structural)
→ Point Load. Set Name = load_top.
    31. Under Properties, tick Force Y. Set at -100.
    32. Under Entities\Nodes, press Add.
33. Under ⑥ select node 3 (*1.5,1,0*) with a (LC), then (RC).
    34. Press OK.

35. Under Boundary Conditions : **Boundary Conditions** select New (Structural)
→ Point Load. Set Name = load_bottom.
    36. Under Properties, tick Force Y. Set at -100.
    37. Under Entities\Nodes, press Add.
38. Under ⑥ select node 2 (*1.5,0,0*) with a (LC), then (RC).
    39. Press OK.

40. Under Boundary Conditions : **Boundary Conditions** select New (Structural)
→ Point Load. Set Name = load_split.
    41. Under Properties, tick Force Y. Set at -50.
    42. Under Entities\Nodes, press Add.
43. Under ⑥ select node 2 (*1.5,0,0*) and 3 (*1.5,1,0*) with a (RC), then (LC).
    44. Press OK.

Defining Loadcases

45. Under Loadcases : **Loadcases** select New → Static.
    46. Set Name = top.
    47. Press Loads. Untick load_bottom, load_split. Press OK.
    48. Set Stepping Procedure\#Steps = 1. Press OK.

49. Under Loadcases : **Loadcases** select New → Static.
    50. Set Name = bottom.
    51. Press Loads. Untick load_top, load_split. Press OK.
    52. Set Stepping Procedure\#Steps = 1. Press OK.

53. Under Loadcases : **Loadcases** select New → Static.
    54. Set Name = split.
    55. Press Loads. Untick load_top, load_bottom. Press OK.
    56. Set Stepping Procedure\#Steps = 1. Press OK.

Running the jobs

Job 1

57. Under Jobs : **Job** select New → Structural. Set Name = top_load.
 58. Under Available, select top.
 59. Press Initial Loads. Untick load_bottom, load_split. Press OK.
 60. Press Check; See in ⑧ if there are any errors.
 61. If there are none, press Run.
   62. In 'Run Job', press Advanced Job Submission.
    63. Press Save Model.
    64. Press Write Input File. Press OK.
    65. Press Submit 1.
   66. Wait until Status = Complete. Press OK
 67. Press OK.

Job 2

68. Under Jobs : **Job** select New → Structural. Set Name = bottom_load.
 69. Under Available, select bottom.
 70. Press Initial Loads. Untick load_top, load_split. Press OK.
 71. Press Check; See in (8) if there are any errors.
 72. If there are none, press Run.
   73. In 'Run Job', press Advanced Job Submission.
    74. Press Save Model.
    75. Press Write Input File. Press OK.
    76. Press Submit 1.
   77. Wait until Status = Complete. Press OK
 78. Press OK.

Job 3

79. Under Jobs : **Job** select New → Structural. Set Name = split_load.
 80. Under Available, select split.
 81. Press Initial Loads. Untick load_top, load_bottom. Press OK.
 82. Press Check; See in ⑧ if there are any errors.
 83. If there are none, press Run.
   84. In 'Run Job', press Advanced Job Submission.
    85. Press Save Model.
    86. Press Write Input File. Press OK.
    87. Press Submit 1.
   88. Wait until Status = Complete.
   89. Press Open Post File (Model Plot Results Menu).

Table 5.3 Summary of numerical results for the simply supported beam problem

Case	Quantity	Node 2	Node 3
Top Loadcase	u_Y	-0.0435789	x.xxxxxxx
	u_X	x.xxxxxxx	x.xxxxxxx
Bottom Loadcase	u_Y	x.xxxxxxx	x.xxxxxxx
	u_X	x.xxxxxxx	x.xxxxxxx
Split Loadcase	u_Y	x.xxxxxxx	x.xxxxxxx
	u_X	x.xxxxxxx	x.xxxxxxx

Viewing the model

90. Under Deformed Shape\Style, select Deformed and Original.
    91. Under Scalar Plot\Style, select Numerics.
        92. Press Scalar and select Displacement Y. Press OK.
        93. Press Scalar and select Displacement X. Press OK.
    94. Press OK.

Open Job beam_plane_bottom_load.t16

95. Under File → Results → Open. Open beam_plane_bottom_load.t16.
    96. Press Scalar and select Displacement Y. Press OK.
    97. Press OK.

Open Job beam_plane_split_load.t16

98. Under File → Results → Open. Open beam_plane_split_load.t16.
    99. Press Scalar and select Displacement X. Press OK.
    100. Press OK.

Results

Collect the results for this problem in Table 5.3.

5.3   Advanced Example

5.3.1   Mesh Refinement—Simply Supported Beam

In order to gather a more accurate result for the nodal displacement of a simply supported beam, we can refine the mesh of the beam given in the previous example 5.2.2. We will subdivide each one of the two elements into a) 4, b) 16 elements. To do so, add these steps after step 7, then continue with step 8 above:

Table 5.4 Summary of numerical results for the mesh refinement problem

	2 Elements	8 Elements	32 Elements
Analytical Bernoulli	x.xxxx		
Analytical Timoshenko	x.xxxx		
Top Loadcase[a]	0.0452456	x.xxxxxxx	x.xxxxxxx
Bottom Loadcase	x.xxxxxxx	x.xxxxxxx	x.xxxxxxx
Split Loadcase	x.xxxxxxx	x.xxxxxxx	x.xxxxxxx

[a] The value for the solution is the average of all deformation values at $x = 2a$

8. Under Geometry & Mesh : **Operations** select Subdivide.
9. Set the first three rows to 2; 2; 1.
10. Press Elements, then in ⑥ select the elements with a (LC), then (RC). Press OK.
11. Under Geometry & Mesh : **Operations** select Sweep.
12. Set Sweep/Tolerance to 0.01.
13. Under Remove Unused, press Nodes. Press OK.

Results

Collect your results for the vertical displacement in the middle of the beam in Table 5.4, including a comparison of the results to the analytical EULER- BERNOULLI and TIMOSHENKO solution.

5.3.2 Stress Concentration

Problem Description

Given is the classical problem of a plate with a circular hole under plane stress conditions, see Fig. 5.6a. The dimensions of the finite plate are $2A \times 2B \times t$ while the hole has a diameter of $2a$. Two opposite edges are loaded by a constant stress σ. The normal stress component σ_Y in the minimum cross section increases towards the root of the hole. This phenomenon is called stress concentration. This stress increase must be considered when the damage or failure of the plate is investigated.

The analytical solution of this problem is given for a semi-infinite plate in spscitePil05 as

$$\frac{\sigma_{max}}{\sigma_{nom}} = 3.0 - 3.140\left(\frac{a}{A}\right)^1 + 3.667\left(\frac{a}{A}\right)^2 - 1.527\left(\frac{a}{A}\right)^3 \quad \text{for } 0 \le \frac{a}{A} \le 1,$$

$$(5.2)$$

Fig. 5.6 Finite plate with a circular hole: **a** entire configuration and **b** under consideration of symmetry

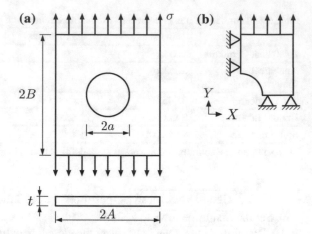

where the nominal stress in the minimum cross section is defined for a plate which is loaded by a force F as:

$$\sigma_{nom} = \frac{F}{2(A - a)t}. \tag{5.3}$$

Consider the symmetry of the problem (see Fig. 5.6b) and calculate the finite element solution of the problem for $2A = 20$, $2B = 20$, $t = 1$, $E = 200000$, $v = 0.3$, and $\sigma = 10$. Compare your result for the stress concentration with the analytical solution.

Marc Solution

Save as 'plate_conc'.

Constructing the mesh

1. Under ⟨ Geometry & Mesh ⟩: **Basic Manipulation** select Geometry and Mesh.
    2. Under Geometry\Curves, select $\overline{\text{Arc Cen/Pnt/Pnt}}$ (see Fig. 5.7).

    3. Under Geometry\Curves, select $\overline{\text{Add}}$.

4. *0,0,0* ENTER *1,0,0* ENTER *0,1,0* ENTER.
    5. Under Geometry\Points, select $\overline{\text{Add}}$.

6. *10,0,0* ENTER *10,10,0* ENTER *0,10,0* ENTER.
    7. Under Geometry\Curves, select $\overline{\text{Polyline}}$.
    8. Under Geometry\Curves, select $\overline{\text{Add}}$.

Fig. 5.7 Setting the curve type

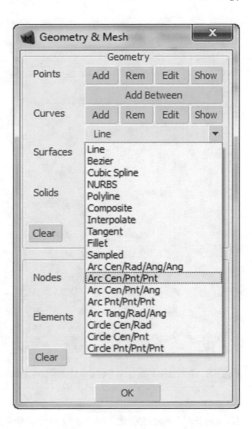

9. In ⑥, click on points 4-6 in counter-clockwise direction with a (LC), then (RC).
  10. Under Geometry\Surfaces, select $\overline{\text{Ruled}}$.
  11. Under Geometry\Surfaces, select $\overline{\text{Add}}$.

12. In ⑥, click on arc 1 and polyline 2 with a (LC), then (RC). The result is shown in Fig. 5.8.
  13. Press $\overline{\text{OK}}$.

14. Under $\boxed{\text{Geometry \& Mesh}}$: **Operations** select Convert.
  15. Set Divisions = 4 4 and press $\overline{\text{Convert}}$.

16. In ⑥, click on surface 1 with a (LC), then (RC). The result is shown in Fig. 5.9. The arrows near the element edges indicate that the elements are oriented clockwise. However, planar elements must be oriented counter-clockwise.
  17. Press $\overline{\text{OK}}$.

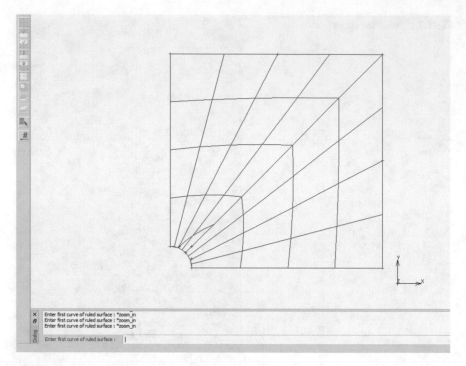

Enter first curve of ruled surface : *zoom_in
Enter first curve of ruled surface : *zoom_in
Enter first curve of ruled surface : *zoom_in

Enter first curve of ruled surface : |

Fig. 5.8 Generated surface 1 of the stress concentration problem

18. Under Geometry & Mesh : **Operations** select Check.
    19. Under Check Elements, select $\overline{\text{Upside Down (2-D)}}$ and $\overline{\text{Flip Elements}}$.
    20. In ⑦, press the first symbol, 'All Existing'.

Setting the Geometric Properties

21. Under Geometric Properties : **Geometric Properties** select New (Structural).
    22. Select → Planar → Plane Stress.
    23. Set Properties\Thickness = 1.
    24. Under Entities\Elements, press $\overline{\text{Add}}$.

25. In ⑦, press the first symbol, 'All Existing'.
    26. Press $\overline{\text{OK}}$.

Setting the Material Properties

27. Under Material Properties : **Material Properties** select New → Finite Stiffness
Region → $\overline{\text{Standard}}$.
    28. Set Other Properties\Young's Modulus = 200000.

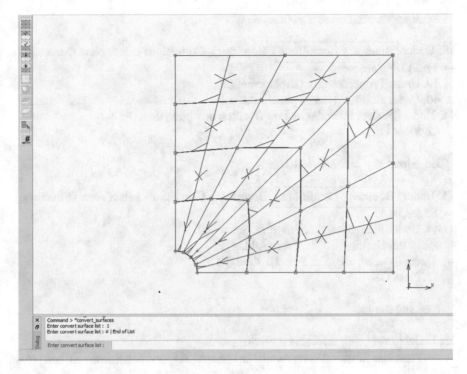

Command > *convert_surfaces
Enter convert surface list : 1
Enter convert surface list : # | End of List

Enter convert surface list :

Fig. 5.9 Generated 4 × 4 mesh of the stress concentration problem

29. Set Other Properties\Poisson's Ratio = 0.3.
30. Under Entities\Elements, press $\overline{\text{Add}}$.

31. In ⑦, press the first symbol, 'All Existing'.
32. Press $\overline{\text{OK}}$.

Setting the Boundary Conditions

Symmetry condition at $x = 0$: $u_X = 0$

33. Under $\boxed{\text{Boundary Conditions}}$: **Boundary Conditions** select New (Structural)
→ Fixed Displacement.
34. Under Properties tick Displacement X.
35. Under Entities\Nodes, press $\overline{\text{Add}}$.
36. Under ⑥ select all nodes at $X = 0$ with a box pick, then (RC).
37. Press $\overline{\text{OK}}$.

Symmetry condition at $y = 0$: $u_Y = 0$

38. Under ⎢Boundary Conditions⎢: **Boundary Conditions** select New (Structural) → Fixed Displacement.
    39. Under Properties tick Displacement Y.
    40. Under Entities\Nodes, press $\overline{\text{Add}}$.
41. Under ⑥ select all nodes at $Y = 0$ with a box pick, then (RC).
    42. Press $\overline{\text{OK}}$.

Edge load at $y = 10$: $\sigma = 10$

43. Under ⎢Boundary Conditions⎢: **Boundary Conditions** select New (Structural) → Edge Load.
    44. Under Properties, tick Pressure. Set at -10.
    45. Under Entities\Edges, press $\overline{\text{Add}}$.
46. Under ⑥ select edges at $Y = 10$, then (RC).
    47. Press $\overline{\text{OK}}$.

Running the Jobs

48. Under ⎢Jobs⎢: **Job** select New → Structural.
    49. Press $\overline{\text{Job Results}}$. Under Available Element Tensors, tick Stress and then press $\overline{\text{OK}}$.
    50. Press $\overline{\text{Check}}$; See in ⑧ if there are any errors.
    51. If there are none, press $\overline{\text{Run}}$.
        52. In 'Run Job', press $\overline{\text{Advanced Job Submission}}$.
            53. Press $\overline{\text{Save Model}}$.
            54. Press $\overline{\text{Write Input File}}$. Press $\overline{\text{OK}}$.
            55. Press $\overline{\text{Submit 1}}$.
        56. Wait until Status = Complete.
        57. Press $\overline{\text{Open Post File}}$ (Model Plot Results Menu).

Viewing the model

58. Under Deformed Shape\Style, select $\overline{\text{Deformed and Original}}$.
    59. Under Scalar Plot\Style, select $\overline{\text{Numerics}}$.
        60. Press $\overline{\text{Scalar}}$ and select Comp 22 of Stress. Check the stress in the root of the hole.
    62. Press $\overline{\text{OK}}$.

Results

MSC Marc: $\sigma_{22} = 18.2835$. Analytical solution: $\sigma_{max} = 30.2349$.

Additional Questions

1. Repeat the simulations for meshes with a division of 8×8, 16×16, and 32×32. Use the 'Subdivide' command to refine the mesh and check for double nodes using the 'Sweep' command[1]. Evaluate the stress intensities again and compare to the analytical value.
2. Create a new 4×4 mesh with Bias Factors of 0 -0.5 and evaluate the stress intensity again. Refine this biased mesh with new divisions of 8×8, 16×16, and 32×32. Evaluate the stress intensities again and compare to the analytical value.
3. Show all the four regular meshes on a single page and on another single page all the four biased meshes.

5.3.3 Stress Intensity/Singularity

Problem Description

Given is the classical problem of a plate with a sharp inner corner under plane stress conditions, see Fig. 5.10. The outer dimensions of the finite plate are $10 \times 10 \times 0.1$ while the cutout has the dimensions 5×5. The top edge is translated by a horizontal displacement of $u_X = -0.05$ and the bottom edge is fixed. The stress components increase towards the inner edge. This phenomenon is called stress intensity or stress singularity. Assume the following material parameters: $E = 200000$ and $\nu = 0.3$. All values are given in consistent units.

Use the equivalent von Mises stress to characterize the stress intensity as a function of the mesh density.

Marc Solution

Save as 'plate_sing'.

Constructing the mesh

1. Under $\boxed{\text{Geometry \& Mesh}}$: **Basic Manipulation** select Geometry and Mesh.
   2. Under Mesh\Nodes, select $\overline{\text{Add}}$.
3. Nodes: 1.$(0,0,0)$ 2.$(10,0,0)$ 3.$(10,5,0)$ 4.$(5,5,0)$ 5.$(0,5,0)$ 6.$(5,10,0)$ 7.$(0,10,0)$.
   4. Under Mesh\Elements, select $\overline{\text{Quad (4)}}$.
   5. $\overline{\text{Add}}$.
6. In ⑥, click on nodes 1,2,3,5 beginning with the bottom left node (1) and continuing the selection in counter-clockwise direction. Then select nodes 5, 4, 6, 7.

---

[1]Set the Sweep Tolerance to 1/10 of the smallest distance between nodes.

Fig. 5.10 Plane stress
singularity problem

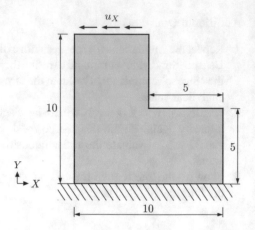

7. Press $\overline{\text{OK}}$.

8. Under | Geometry & Mesh | : **Operations** select Subdivide.
    9. Set Divisions = 4 2 1 and press $\overline{\text{Elements}}$.

10. In ⑥, click on element 1 with a (LC), then (RC).
    11. Set Divisions = 2 2 1 and press $\overline{\text{Elements}}$.

12. In ⑥, click on element 2 with a (LC), then (RC). The result is shown in Fig. 5.11.
The arrows near the element edges indicate that the elements are oriented counter-
clockwise, i.e. the required orientation for plane elements.
    13. Press $\overline{\text{OK}}$.

14. Under | Geometry & Mesh | : **Operations** select Sweep.
    15. Set Tolerance[2] = 0.25 and press $\overline{\text{all}}$.
    16. Press $\overline{\text{OK}}$.

Setting the Geometric Properties

17. Under | Geometric Properties | : **Geometric Properties** select New (Structural).
    18. Select → Planar → Plane Stress.
    19. Set Properties\Thickness = 0.1
    20. Under Entities\Elements, press $\overline{\text{Add}}$.

21. In ⑦, press the first symbol, 'All Existing'.
    22. Press $\overline{\text{OK}}$.

---

[2]As a rule of thumb, one may take one-tenth of the smallest distance between two nodes.

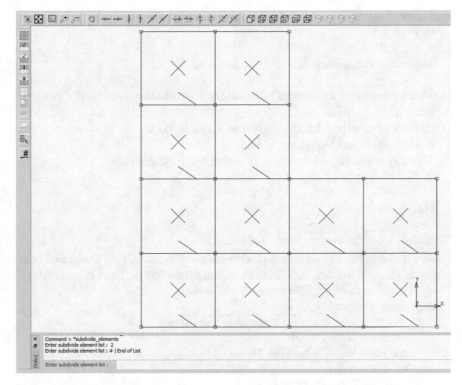

Fig. 5.11 Generated 4×4 mesh of the stress singularity problem

Setting the Material Properties

23. Under Material Properties : **Material Properties** select New → Finite Stiffness Region → Standard.
    24. Set Other Properties\Young's Modulus = 200000.
    25. Set Other Properties\Poisson's Ratio = 0.3.
    26. Under Entities\Elements, press Add.

27. In ⑦, press the first symbol, 'All Existing'.
    28. Press OK.

Setting the Boundary Conditions

Fixed Support at $Y = 0$:

29. Under Boundary Conditions : **Boundary Conditions** select New (Structural) → Fixed Displacement.
    30. Under Properties tick Displacement X and Displacement Y.
    31. Under Entities\Nodes, press Add.

32. Under ⑥ select all nodes at $Y = 0$ with a box pick, then (RC).
   33. Press $\overline{\text{OK}}$.

Displacement condition at $Y = 10$: $u_X = -0.05$

34. Under $\boxed{\text{Boundary Conditions}}$: **Boundary Conditions** select New (Structural)
→ Fixed Displacement.
   35. Under Properties tick Displacement X. Set at -0.05
   36. Under Entities\Nodes, press $\overline{\text{Add}}$.
37. Under ⑥ select all nodes at $Y = 10$ with a box pick, then (RC).
   38. Press $\overline{\text{OK}}$.

Running the Jobs

39. Under $\boxed{\text{Jobs}}$: **Job** select New → Structural.
   40. Press $\overline{\text{Job Results}}$. Under Available Element Tensors, tick Stress and under
Available Element Scalars, tick Equivalent Von Mises Stress. Then press $\overline{\text{OK}}$.
   41. Press $\overline{\text{Check}}$; See in ⑧ if there are any errors.
   42. If there are none, press $\overline{\text{Run}}$.
      43. In 'Run Job', press $\overline{\text{Advanced Job Submission}}$.
         44. Press $\overline{\text{Save Model}}$.
         45. Press $\overline{\text{Write Input File}}$. Press $\overline{\text{OK}}$.
         46. Press $\overline{\text{Submit 1}}$.
      47. Wait until Status = Complete.
      48. Press $\overline{\text{Open Post File}}$ (Model Plot Results Menu).

Viewing the model

49. Under Deformed Shape\Style, select $\overline{\text{Deformed and Original}}$.
   50. Under Scalar Plot\Style, select $\overline{\text{Numerics}}$.
      51. Press $\overline{\text{Scalar}}$ and select Equivalent Von Mises Stress. Check the stress in
the inner corner of the plate.
      52. Press $\overline{\text{OK}}$.

Results

MSC Marc: $\sigma_{\text{Mises}} = 571.618$.

Additional Questions

1. Repeat the simulations for meshes with a division of 8×8, 16×16, and 32×32.
   Use the Subdivide command to refine the mesh and check for double nodes using
   the Sweep command[3]. Evaluate the von Mises stress at the inner corner again.

---

[3] Set the Sweep Tolerance to 1/10 of the smallest distance between nodes.

2. Create a new 4×4 mesh with Bias Factors of $|0.5|$ $|0.5|$ and evaluate again the von Mises stress at the inner corner. Refine this biased mesh with new divisions of 8×8, 16×16, and 32×32. Evaluate again the von Mises stress at the inner corner. Why are the stress values in both cases, i.e., regular and biased meshes, not converging?
3. Show on a single page all the four regular meshes and on another single page all the four biased meshes.

5.3.4 Short Fiber Reinforced Composite Plate

Problem Description

Given is a short fiber reinforced composite plate with aligned fibers as shown in Fig. 5.12. The outer dimensions of the plate are $200 \times 200 \times 2$ (mm) and a plane stress state is assumed. The matrix material (index 'm') is epoxy (linear-elastic properties: $E_m = 3000$ MPa, $\nu_m = 0.35$). The short fibers (index 'f') are made of E-glass with a constant length of $a = 5$ (mm), constant cross section $A_f = 0.1$ mm^2, and pure linear-elastic behavior ($E_f = 40000$ MPa).

The task is to predict the macroscopic stiffness of the composite plate for a given fiber volume fraction ϕ_f. The following procedure outlines a stepped process to achieve this task.

The first step consists in considering the matrix (two-dimensional plane stress quad 4 element) and the fibers (one-dimensional line 2 rod element) separately under a tensile test situation, see Fig. 5.13. Apply on both models a constant boundary

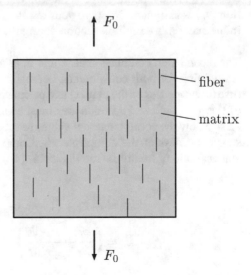

Fig. 5.12 Tensile loading of a composite plate

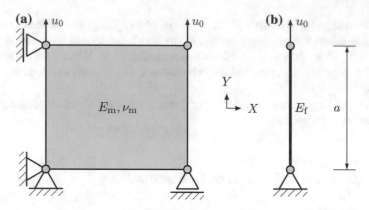

Fig. 5.13 Computational single-element model for tensile testing: **a** matrix and **b** fiber

displacement of $u_0 = 0.1$ (mm) parallel to the fibers, i.e. in direction of the global Y-axis.

The macroscopic stiffness (YOUNG's modulus) can be calculated as follows:

$$E_m = \frac{\sigma_m}{\varepsilon_m} = \frac{\frac{\sum_{i=1}^{2} F_{Ri}}{A_m}}{\frac{u_0}{a}}, \tag{5.4}$$

$$E_f = \frac{\sigma_f}{\varepsilon_f} = \frac{\frac{F_R}{A_f}}{\frac{u_0}{a}}, \tag{5.5}$$

where F_R is the reaction force at those nodes where a displacement boundary condition (u_0) is assigned. Compare your results with the defined material properties (E) in the material's definition section during the model generation.

The second step combines matrix and fiber to a composite material, see Fig. 5.14. Assign first for both constituents, i.e. matrix and fiber, the same elastic properties (use the properties of the matrix) and calculate the macroscopic stiffness in the sense of Eqs. (5.4) and (5.5). Which area must be used to obtain the correct value?
Assign now the correct material values for fiber and matrix and calculate the macroscopic stiffness of the composite E_c. Compare this finite element result with the classical rule of mixtures (see Fig. 5.15) [4]:

$$E_c = \phi_f E_f + (1 - \phi_f) E_m \quad \text{(upper bound)}, \tag{5.6}$$

$$E_c = \left(\frac{\phi_f}{E_f} + \frac{1 - \phi_f}{E_m} \right)^{-1} \quad \text{(lower bound)}, \tag{5.7}$$

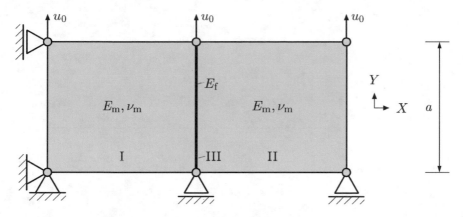

Fig. 5.14 Computational three-element model for tensile testing of a composite material

Fig. 5.15 Rule of mixtures

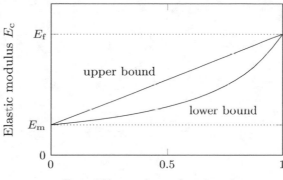

where the fiber volume fraction is the following volume ratio:

$$\phi_f = \frac{V_f}{V_c} = \frac{V_f}{V_m + V_f}. \qquad (5.8)$$

The final step considers a plate of size $100 \times 100 \times 2$ (mm) under consideration of symmetry conditions. Create a single matrix element of this size and subdivide it into 20 equal steps in the X- and Y-direction. This results in smaller matrix elements of edge length a. Now randomly distribute 40 fiber elements parallel to the global Y-direction and simulate a tensile test as outlined above. Evaluate the macroscopic stiffness and compare this finite element result with the rule of mixtures. For illustration purposes, it is best to present a figure of the matrix elements and a figure of the fiber elements separately.

Chapter 6
Classical Plate Elements

The definition of a thin or KIRCHHOFF plate element is summarized in Table 6.1. The derivation in lectures normally starts with the introduction of an elemental coordinate system (x, y, z) where the x- and y-axis is aligned with the principal axes of the element and the z-axis is perpendicular to the element. Based on the definition of this element, displacements $(u_{1z}, u_{2z}, u_{3z}, u_{4z})$ can only occur perpendicular to the plane while the rotations $(\varphi_{ix}, \varphi_{iy})$ $(i = 1, \ldots, 4)$ act around the x- and y-axis.

The implementation of a plane element in a commercial finite element code is more general, i.e. based on the global coordinate system (X, Y, Z). Thus, the geometry is defined based on the global coordinates of each node (X_i, Y_i, Z_i) and the thickness h. Furthermore, such an element is similar to a generalized beam, i.e. the superposition of a beam with a rod, or precisely here the superposition of a classical plate with a plane elasticity element.

MSC Marc element type 139 is a thin-shell element which uses bilinear interpolation for the coordinates, the displacements, and the rotations. In regards to degrees of freedom, strains and stresses, the following outputs are obtained:

- Degrees of freedom obtained in global coordinate system X, Y, Z:
  Three displacements per node i: u_{iX}, u_{iY}, u_{iZ}
  Three rotations per node i: $\varphi_{iX}, \varphi_{iY}, \varphi_{iZ}$

- Generalized strains in local coordinate system:
  Middle surface stretches: $\varepsilon_x, \varepsilon_y, \varepsilon_x$
  Middle surface curvatures: $\kappa_x, \kappa_y, \kappa_x$

- Stresses in local coordinate system:
  $\sigma_x, \sigma_y, \sigma_x$ at equally spaced layers through thickness

A. Öchsner and M. Öchsner, *A First Introduction to the Finite Element Analysis Program MSC Marc/Mentat*, https://doi.org/10.1007/978-3-319-71915-3_6

Table 6.1 Definition of plate elements

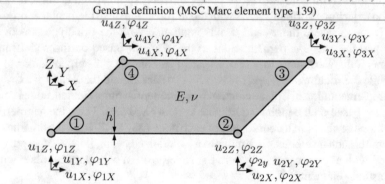

Definitions	Degrees of freedom
Material: E, ν	Displacements: $u_{1z}, u_{2z}, u_{3z}, u_{4z}$
Geometry: a, b, h	Rotations:
	$\varphi_{1x}, \varphi_{2x}, \varphi_{3x}, \varphi_{xy}, \varphi_{1y}, \varphi_{2y}, \varphi_{3y}, \varphi_{4y}$

General definition (MSC Marc element type 139)

Definitions	Degrees of freedom
Material: E, ν	Node 1: $u_{1X}, u_{1Y}, u_{1Z}, \varphi_{1X}, \varphi_{1Y}, \varphi_{1Z}$
Geometry: h	Node 2: $u_{2X}, u_{2Y}, u_{2Z}, \varphi_{2X}, \varphi_{2Y}, \varphi_{2Z}$
Node 1: X_1, Y_1, Z_1	Node 3: $u_{3X}, u_{3Y}, u_{3Z}, \varphi_{3X}, \varphi_{3Y}, \varphi_{3Z}$
Node 2: X_2, Y_2, Z_2	Node 4: $u_{4X}, u_{4Y}, u_{4Z}, \varphi_{4X}, \varphi_{4Y}, \varphi_{4Z}$
Node 3: X_3, Y_3, Z_3	
Node 4: X_4, Y_4, Z_4	

6.1 Basic Examples

6.1.1 Plate Element Under Bending Load

Problem Description

Given is a regular two-dimensional thin plate element as shown in Fig. 6.1, with a length of $2a$, a width of $2b$ and a thickness of h. The left-hand nodes are fixed and the right-hand nodes are loaded by vertical point loads $F_z = -100$. Use a single plate element to calculate the nodal deformations for $a = 0.75$, $b = 0.5$, $h = 0.05$, $E = 200000$, and $\nu = 0.2$.

Marc Solution

Save as 'plate_canti'.

Constructing the mesh

1. Under Geometry & Mesh : **Basic Manipulation** select Geometry and Mesh (see Fig. 2.3).
    2. Under Mesh\Nodes, select $\overline{\text{Add}}$.
3. *0,0,0* ENTER *1.5,0,0* ENTER *1.5,1,0* ENTER *0,1,0* ENTER.
    4. Under Mesh\Elements, select $\overline{\text{Quad (4)}}$ (see Fig. 5.2).
    5. $\overline{\text{Add}}$.
6. In ⑥, click on nodes 1–4, beginning with the bottom left node (1) and continuing the selection in counter-clockwise direction.
    7. Press $\overline{\text{OK}}$.

Fig. 6.1 Schematic drawing of the plate under bending load

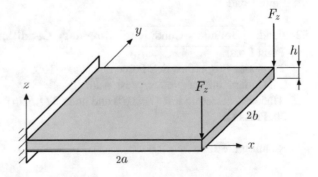

Setting the Geometric Properties

8. Under Geometric Properties : **Geometric Properties** select New (Structural).
   9. Select → 3-D → Shell.
   10. Set Properties\Thickness = 0.05.
   12. Under Entities\Elements, press $\overline{\text{Add}}$.
13. In ⑥ select the element with a (LC), then (RC).
   14. Press $\overline{\text{OK}}$.

Setting the Material Properties

15. Under Material Properties : **Material Properties** select New → Finite Stiffness
Region → Standard.
   16. Set Other Properties\Young's Modulus = 200000 (see Fig. 2.6).
   17. Set Other Properties\Poisson's Ratio = 0.2.
   18. Under Entities\Elements, press $\overline{\text{Add}}$.
19. In ⑥ select the element with a (LC), then (RC).
   20. Press $\overline{\text{OK}}$.

Setting the Boundary Conditions

Support Conditions

21. Under Boundary Conditions : **Boundary Conditions** select New (Structural)
→ Fixed Displacement (see Fig. 2.7).
   22. Under Properties tick Displacement X, Y, Z and Rotation X, Y, Z.
   23. Under Entities\Nodes, press $\overline{\text{Add}}$.
24. Under ⑥ select node 1 (*0,0,0*) and node 4 (*0,1,0*) with a (LC), then (RC).
   25. Press $\overline{\text{OK}}$.

Point Load

26. Under Boundary Conditions : **Boundary Conditions** select New (Structural)
→ Point Load.
   27. Under Properties, tick Force Z. Set at −100.
   28. Under Entities\Nodes, press $\overline{\text{Add}}$.
29. Under ⑥ select node 2 (*1.5,0,0*) and node 3 (*1.5,1,0*) with a (LC), then (RC).
   30. Press $\overline{\text{OK}}$.

Running the Jobs

31. Under Jobs : **Element Types** select Element Types.
   32. Under Analysis Dimension, select 3-D.

Table 6.2 Summary of numerical results for the cantilever plate bending problem

Quantity	Node 2	Node 3
u_z	-104.712	x.xxxxxx
φ_y	105.744	x.xxxxxx
φ_x	x.xxxxxx	x.xxxxxx

33. Press $\overline{\text{Shell/Membrane}}$, then press $\overline{139}$.
34. In ⑦, press the first symbol, 'All Existing'.
   35. Press $\overline{\text{OK}}$. Press $\overline{\text{OK}}$.
36. Under $\boxed{\text{Jobs}}$: **Job** select New → Structural.
   37. Press $\overline{\text{Check}}$; See in ⑧ if there are any errors.
   38. If there are none, press $\overline{\text{Run}}$.
      39. In 'Run Job', press $\overline{\text{Advanced Job Submission}}$.
         40. Press $\overline{\text{Save Model}}$.
         41. Press $\overline{\text{Write Input File}}$. Press $\overline{\text{OK}}$.
         42. Press $\overline{\text{Submit 1}}$.
      43. Wait until Status = Complete.
      44. Press $\overline{\text{Open Post File (Model Plot Results Menu)}}$.

   Viewing the model

45. Under Deformed Shape\Style, select $\underline{\text{Original}}$.
   46. Under Scalar Plot\Style, select $\overline{\text{Numerics}}$.
      47. Press $\overline{\text{Scalar}}$ and select Displacement Z. Press $\overline{\text{OK}}$.
   48. Press $\overline{\text{OK}}$.

Results

Collect the results for this problem in Table 6.2.
Compare the results from Table 6.2 with the analytical solution from the EULER-BERNOULLI beam theory.

6.1.2 Simply Supported Plate Element

Problem Description

Given is a regular two-dimensional thin plate element as shown in Fig. 6.2, with a length of $2a$, a width of $2b$ and a thickness of h. All four corners are simply supported and the center of the plate is loaded by a vertical point load $F_z = -100$. Use four plate elements to calculate the deflection in the middle of the plate (i.e., for $x = 0.75$)

Fig. 6.2 Schematic drawing
of the simply supported plate
under bending load

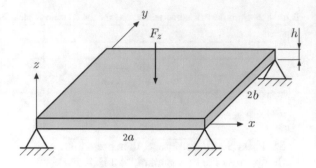

for $a = 0.75$, $b = 0.5$, $h = 0.05$, $E = 200000$, and $\nu = 0.2$.

Marc Solution

Save as 'plate_simple'.

Constructing the mesh

1. Under Geometry & Mesh : **Basic Manipulation** select Geometry and Mesh (see Fig. 2.3).
   2. Under Mesh\Nodes, select Add.
3. *0,0,0* ENTER *1.5,0,0* ENTER *1.5,1,0* ENTER *0,1,0* ENTER.
   4. Under Mesh\Elements, select Quad (4) (see Fig. 5.2).
   5. Add.
6. In ⑥, click on nodes 1–4, beginning with the bottom left node (1) and continuing the selection in counter-clockwise direction.
   7. Press OK.
8. Under Geometry & Mesh : **Operations** select Subdivide.
   9. Press Divisions.
10. *2,2,0* ENTER.
    11. Press Elements.
12. In ⑥ select the element with a (LC), then (RC).
    13. Press OK.
14. Under Geometry & Mesh : **Operations** select Sweep.
    15. Under Sweep, set Tolerance at 0.05.
    16. Under Sweep, press All.
    17. Press OK.

Setting the Geometric Properties

18. Under Geometric Properties : **Geometric Properties** select New (Structural).
    19. Select → 3-D → Shell.

20. Set Properties\Thickness = 0.05.
21. Under Entities\Elements, press $\overline{\text{Add}}$.
22. In ⑥ select the four elements with a (LC), then (RC).
    23. Press $\overline{\text{OK}}$.

Setting the Material Properties

24. Under | Material Properties |: **Material Properties** select New → Finite Stiffness Region → Standard.
    25. Set Other Properties\Young's Modulus = 200000 (see Fig. 2.6).
    26. Set Other Properties\Poisson's Ratio = 0.2.
    27. Under Entities\Elements, press $\overline{\text{Add}}$.
28. In ⑥ select the four elements with a (LC), then (RC).
    29. Press $\overline{\text{OK}}$.

Setting the Boundary Conditions

Support Conditions

30. Under | Boundary Conditions |: **Boundary Conditions** select New (Structural) → Fixed Displacement (see Fig. 2.7).
    31. Under Properties tick Displacement X, Y, and Z.
    32. Under Entities\Nodes, press $\overline{\text{Add}}$.
33. Under ⑥ select the four corner nodes (*0,0,0*), (*1.5,0,0*), (*1.5,1,0*) and (*0,1,0*) with a (LC), then (RC).
    34. Press $\overline{\text{OK}}$.

Point Load

35. Under | Boundary Conditions |: **Boundary Conditions** select New (Structural) → Point Load.
    36. Under Properties, tick Force Z. Set at −100.
    37. Under Entities\Nodes, press $\overline{\text{Add}}$.
38. Under ⑥ select the center node (*0.75,0.5,0*) with a (LC), then (RC).
    39. Press $\overline{\text{OK}}$.

Running the Jobs

40. Under | Jobs |: **Element Types** select Element Types.
    41. Under Analysis Dimension, select 3-D.
    42. Press Shell/Membrane, then press $\overline{139}$.
43. In ⑦, press the first symbol, 'All Existing'.
    44. Press $\overline{\text{OK}}$. Press $\overline{\text{OK}}$.

Table 6.3 Summary of numerical results for the simply supported plate bending problem

Quantity	(0.75, 0)	(0.75, 0.5)	(0.75, 1.0)
u_z	−3.18921	x.xxxxxx	x.xxxxxx

45. Under | Jobs |: **Job** select New → Structural.
   46. Press Check; See in ⑧ if there are any errors.
     47. If there are none, press Run.
       48. In 'Run Job', press Advanced Job Submission.
         49. Press Save Model.
         50. Press Write Input File. Press OK.
         51. Press Submit 1.
       52. Wait until Status = Complete.
       53. Press Open Post File (Model Plot Results Menu).

Viewing the model

54. Under Deformed Shape\Style, select Original.
   55. Under Scalar Plot\Style, select Numerics.
     56. Press Scalar and select Displacement Z. Press OK.
   57. Press OK.

Results

Collect the results for this problem in Table 6.3.

Compare the results from Table 6.3 with the analytical solution from the EULER-BERNOULLI beam theory.

6.2 Advanced Example

6.2.1 Mesh Refinement—Simply Supported Plate

In order to gather a more accurate result for the nodal displacement in the middle of the simply supported plate, we can refine the mesh of the plate given in the previous example in Sect. 6.1.2. We will subdivide the original configuration of 4 elements into (a) 16 and (b) 64 elements.

Results

The following vertical displacements are obtained in the middle of the plate:
16 elements: −3.75916;
64 elements: −3.80849.

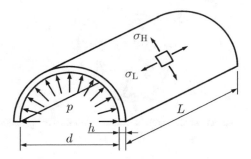

Fig. 6.3 Half of a thin cylinder which is loaded by an internal pressure

6.2.2 Stresses in Thin Cylinders and Shells (Pressure Vessels)

The following section summarizes a few basic results in regards to the stress state in *thin* cylinders and shells spsciteTim40,Pur12. Thin in this context means that the ratio between the radial thickness h and the inner diameter d is smaller than one-tenth:

$$\frac{h}{d} \leq \frac{1}{10}. \tag{6.1}$$

Thin cylinders and shells are widely used as pressure vessels used for gas storage in the form of cylindrical or spherical tanks.

6.2.3 Pressure Vessel: Thin Cylinder

The stress state in a thin cylindrical pressure vessel loaded by an internal pressure p (see Fig. 6.3) is approximately described by the hoop or circumferential stress

$$\sigma_{\mathrm{H}} = \frac{pd}{2h}, \tag{6.2}$$

and the longitudinal stress

$$\sigma_{\mathrm{L}} = \frac{pd}{4h}. \tag{6.3}$$

Derive Eqs. (6.2) and (6.3) to understand the assumptions. Develop based on this knowledge a simple finite element model (thin plate elements) to verify these simply

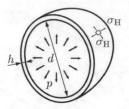

Fig. 6.4 Half of a thin spherical shell which is loaded by an internal pressure

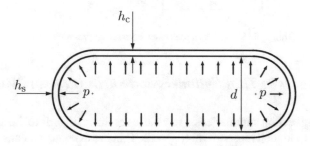

Fig. 6.5 Schematic representation of a thin cylinder with hemispherical ends

analytical relationships. Evaluate the error between the analytical and finite element solution. Assume the following values with consistent units: $L = 500$, $d = 100$, $h = 1$, and material properties $E = 210000$, $v = 0.3$.

6.2.4 Pressure Vessel: Spherical Shell

The stress state in a thin spherical shell loaded by an internal pressure p (see Fig. 6.4) is approximately described by two mutually perpendicular circumferential stresses of equal value:

$$\sigma_{\mathrm{H}} = \frac{pd}{4h}. \tag{6.4}$$

6.2.5 Pressure Vessel: Thin Cylinder with Hemispherical Ends

The stress state in a thin cylinder with hemispherical ends loaded by an internal pressure p (see Fig. 6.5) is approximately described in the cylindrical part by

$$\sigma_{H_c} = \frac{pd}{2h_c}, \quad \sigma_{L_c} = \frac{pd}{4h_c} \tag{6.5}$$

and in the hemispherical ends by

$$\sigma_{H_s} = \frac{pd}{4h_s}. \tag{6.6}$$

Chapter 7
Three-Dimensional Elements

7.1 Definition of Three-Dimensional Elements

The definition of a solid element (hexahedral) is summarized in Table 7.1. This eight-node, isoparametric hexahedral uses trilinear interpolation functions and the stiffness matrix is formed using eight-point Gaussian integration. The node numbering must follow the scheme shown in Table 7.1.

The six strains $(\varepsilon_X, \varepsilon_Y, \varepsilon_Z, \gamma_{XY}, \gamma_{YZ}, \gamma_{ZX})$ and the six stresses $(\sigma_X, \sigma_Y, \sigma_Z, \sigma_{XY}, \sigma_{YZ}, \sigma_{ZX})$ are evaluated at the eight integration points (or the centroid of the element).

7.2 Basic Examples

7.2.1 Solid Under Tensile Load

Problem Description

Given is a regular three-dimensional solid element as shown in Fig. 7.1, with a width of $2a$, a height of $2b$ and a thickness of $2c$. The left-hand nodes are fixed and the right-hand nodes are loaded by a horizontal point load $F = 50$. Use a single hexahedral element to calculate the nodal displacements for $a = 0.75$, $b = 0.5$, $2c = 0.5$, $E = 200000$, and $\nu = 0.2$.

Marc Solution

Save as 'solid_tensile'.

Constructing the mesh

© Springer International Publishing AG 2018
A. Öchsner and M. Öchsner, *A First Introduction to the Finite Element Analysis Program MSC Marc/Mentat*, https://doi.org/10.1007/978-3-319-71915-3_7

Table 7.1 Definition of a three-dimensional hexahedral element

General definition (MSC Marc element type 7)

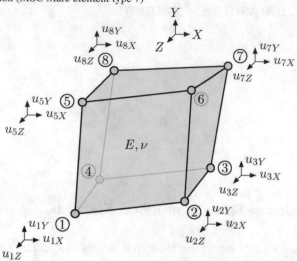

Definitions	Degrees of freedom
Material: E, ν	Node 1: u_{1X}, u_{1Y}, u_{1Z}
Node 1: X_1, Y_1, Z_1	Node 2: u_{2X}, u_{2Y}, u_{2Z}
Node 2: X_2, Y_2, Z_2	Node 3: u_{3X}, u_{3Y}, u_{3Z}
Node 3: X_3, Y_3, Z_3	Node 4: u_{4X}, u_{4Y}, u_{4Z}
Node 4: X_4, Y_4, Z_4	Node 5: u_{5X}, u_{5Y}, u_{5Z}
Node 5: X_5, Y_5, Z_5	Node 6: u_{6X}, u_{6Y}, u_{6Z}
Node 6: X_6, Y_6, Z_6	Node 7: u_{7X}, u_{7Y}, u_{7Z}
Node 7: X_7, Y_7, Z_7	Node 8: u_{8X}, u_{8Y}, u_{8Z}
Node 8: X_8, Y_8, Z_8	

Fig. 7.1 Schematic drawing of the simple solid problem

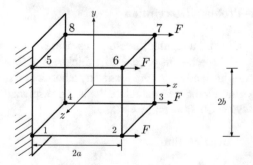

1. Under $\boxed{\text{Geometry \& Mesh}}$: **Basic Manipulation** select Geometry and Mesh (see Fig. 2.3).

  2. Under Mesh\Nodes, select $\overline{\text{Add}}$.

3. Nodes:

Node nr.	Coordinates	Node nr.	Coordinates
1	*(0,0,0)*	5	*(0,1,0)*
2	*(1.5,0,0)*	6	*(1.5,1,0)*
3	*(1.5,0,−1)*	7	*(1.5,1,−1)*
4	*(0,0,-1)*	8	*(0,1,-1)*

  4. Under Mesh\Elements, select $\overline{\text{Hex (8)}}$.
  5. $\overline{\text{Add}}$.
6. In ⑥, click on nodes 1–8, in that order.
  7. Press $\overline{\text{OK}}$.

Setting the Geometric Properties

8. Under $\boxed{\text{Geometric Properties}}$: **Geometric Properties** select New (Structural).
  9. Select 3-D → Solid.
  10. Under Entities\Elements, press $\overline{\text{Add}}$.
11. In ⑥ select the element with a (LC), then (RC).
  12. Press $\overline{\text{OK}}$.

Setting the Material Properties

13. Under $\boxed{\text{Material Properties}}$: **Material Properties** select New → Finite Stiffness Region → Standard.
  14. Set Other Properties\Young's Modulus = 200000 (see Fig. 2.6).
  15. Set Other Properties\Poisson's Ratio = 0.2.
  16. Under Entities\Elements, press $\overline{\text{Add}}$.
17. In ⑥ select the element with a (LC), then (RC).
  18. Press $\overline{\text{OK}}$.

Setting the Boundary Conditions

Support Conditions

19. Under $\boxed{\text{Boundary Conditions}}$: **Boundary Conditions** select New (Structural) → Fixed Displacement (see Fig. 2.7).
  20. Under Properties tick Displacement X, Y and Z.
  21. Under Entities\Nodes, press $\overline{\text{Add}}$.
22. Under ⑥ select nodes 1, 4, 5 and 8 with a (LC), then (RC).

23. Press $\overline{\text{OK}}$.

Point Load

24. Under $\boxed{\text{Boundary Conditions}}$: **Boundary Conditions** select New (Structural) → Point Load.
    25. Under Properties, tick Force X. Set at 50.
    26. Under Entities\Nodes, press $\overline{\text{Add}}$.
27. Under ⑥ select node 2, 3, 6 and 7 with a (LC), then (RC).
    28. Press $\overline{\text{OK}}$.

Running the Jobs

29. Under $\boxed{\text{Jobs}}$: **Job** select New → Structural.
    30. Press $\overline{\text{Check}}$; See in ⑧ if there are any errors.
    31. If there are none, press $\overline{\text{Run}}$.
        32. In 'Run Job', press $\overline{\text{Advanced Job Submission}}$.
            33. Press $\overline{\text{Save Model}}$.
            34. Press $\overline{\text{Write Input File}}$. Press $\overline{\text{OK}}$.
            35. Press $\overline{\text{Submit 1}}$.
        36. Wait until Status = $\overline{\text{Complete}}$.
        37. Press $\overline{\text{Open Post File}}$ (Model Plot Results Menu).

Viewing the model

38. Under Deformed Shape\Style, select $\overline{\text{Deformed and Original}}$.
    39. Under Scalar Plot\Style, select $\overline{\text{Numerics}}$.
        40. Press $\overline{\text{Scalar}}$ and select Displacement Y. Press $\overline{\text{OK}}$.
        41. Press $\overline{\text{Scalar}}$ and select Displacement X. Press $\overline{\text{OK}}$.
    42. Press $\overline{\text{OK}}$.

Results

Collect the results for this problem in Table 7.2.

Table 7.2 Summary of numerical results for the simple solid problem

Quantity	Node 2	Node 3	Node 6	Node 7
u_X	0.00145565	x.xxxxxxxx	x.xxxxxxxx	x.xxxxxxxx
u_Y	x.xxxxxxxx	x.xxxxxxxx	x.xxxxxxxx	x.xxxxxxxx
u_Z	x.xxxxxxxx	x.xxxxxxxx	x.xxxxxxxx	x.xxxxxxxx

7.2.2 Simply Supported Solid

Problem Description

Given is a simply supported beam as indicated in Fig. 7.2. The length of the beam is $4a$ and the rectangular cross section has the dimensions $2b \times 2c$. The beam is loaded by a point force of $F = 200$ acting in the middle of the beam. Use three-dimensional elements in the following to model the problem and to calculate the nodal displacements in the middle of the beam. The modelling approach is based on two elements with nodes 1, ..., 12, see Fig. 7.2. Consider $a = \frac{3}{4}$, $b = \frac{1}{2}$, $c = \frac{1}{2}$ and $E = 200000$, $\nu = 0.2$.

Marc Solution

Save as 'supported_solid'.

Constructing the mesh

1. Under $\boxed{\text{Geometry \& Mesh}}$: **Basic Manipulation** select Geometry and Mesh (see Fig. ??).
  2. Under Mesh\Nodes, select $\overline{\text{Add}}$.

3. Nodes:
  4. Under Mesh\Elements, select $\overline{\text{Hex (8)}}$.
  5. $\overline{\text{Add}}$.
6. In ⑥, click on nodes 1–8, in that order, and then on nodes 2, 9, 10, 3 and 6, 11, 12, 7 in that order.

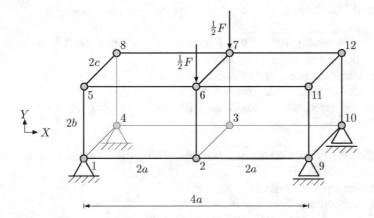

Fig. 7.2 Schematic drawing of the problem

Node nr.	Coordinates	Node nr.	Coordinates
1	(0,0,0)	7	(1.5,1,−1)
2	(1.5,0,0)	8	(0,1,−1)
3	(1.5,0,−1)	9	(3,0,0)
4	(0,0,−1)	10	(3,0,−1)
5	(0,1,0)	11	(3,1,0)
6	(1.5,1,0)	12	(3,1,−1)

7. Press $\overline{OK}$.

Setting the Geometric Properties

8. Under Geometric Properties : **Geometric Properties** select New (Structural).
    9. Select 3-D → Solid.
    10. Under Entities\Elements, press $\overline{Add}$.
11. In ⑥ select the elements with a (LC), then (RC).
    12. Press $\overline{OK}$.

Setting the Material Properties

13. Under Material Properties : **Material Properties** select New → Finite Stiffness Region → Standard.
    14. Set Other Properties\Young's Modulus = 200000 (see Fig. 2.6).
    15. Set Other Properties\Poisson's Ratio = 0.2.
    16. Under Entities\Elements, press $\overline{Add}$.
17. In ⑥ select the elements with a (LC), then (RC).
    18. Press $\overline{OK}$.

Setting the Boundary Conditions

Support Conditions

19. Under Boundary Conditions : **Boundary Conditions** select New (Structural) → Fixed Displacement (see Fig. 2.7).
    20. Under Properties tick Displacement X, Y and Z.
    21. Under Entities\Nodes, press $\overline{Add}$.
22. Under ⑥ select nodes 1 (0,0,0) and 4 (0,0,−1) with a (LC), then (RC).
    23. Press $\overline{OK}$.

24. Under Boundary Conditions : **Boundary Conditions** select New (Structural) → Fixed Displacement.
    25. Under Properties tick Displacement Y.
    26. Under Entities\Nodes, press $\overline{Add}$.
27. Under ⑥ select nodes 9 (3,0,0) and 10 (3,0,−1) with a (LC), then (RC).

28. Press $\overline{\text{OK}}$.

Point Loads

29. Under | Boundary Conditions |: **Boundary Conditions** select New (Structural)
$\rightarrow$ Point Load.
    30. Under Properties, tick Force Y. Set at -100.
    31. Under Entities\Nodes, press $\overline{\text{Add}}$.
32. Under ⑥ select nodes 6 $(1.5,1,0)$ and 7 $(1.5,1,-1)$ with a (LC), then (RC).
    33. Press $\overline{\text{OK}}$.

Running the Jobs

34. Under | Jobs |: **Job** select New $\rightarrow$ Structural.
    35. Press $\overline{\text{Check}}$; See in ⑧ if there are any errors.
    36. If there are none, press $\overline{\text{Run}}$.
        37. In 'Run Job', press $\overline{\text{Advanced Job Submission}}$.
            38. Press $\overline{\text{Save Model}}$.
            39. Press $\overline{\text{Write Input File}}$. Press $\overline{\text{OK}}$.
            40. Press $\overline{\text{Submit 1}}$.
        41. Wait until Status $=$ Complete.
        42. Press $\overline{\text{Open Post File}}$ (Model Plot Results Menu).

Viewing the model

43. Under Deformed Shape\Style, select $\overline{\text{Deformed and Original}}$.
    44. Under Scalar Plot\Style, select $\overline{\text{Numerics}}$.
        45. Press $\overline{\text{Scalar}}$ and select Displacement Y. Press $\overline{\text{OK}}$.
        46. Press $\overline{\text{Scalar}}$ and select Displacement X. Press $\overline{\text{OK}}$.
    47. Press $\overline{\text{OK}}$.

Results

Collect the results for this problem in Table 7.3.

Table 7.3 Summary of numerical results for the simply supported beam

Node	u_X	u_Y	u_Z
2	0.0017768	-0.00433298	-8.56874×10^{-5}
3	X.XXXXXXXX	X.XXXXXXXX	X.XXXXXXXX
6	X.XXXXXXXX	X.XXXXXXXX	X.XXXXXXXX
7	X.XXXXXXXX	X.XXXXXXXX	X.XXXXXXXX

7.3   Advanced Example

7.3.1   Mesh Refinement

In order to gather a more accurate result for the nodal displacement of a simply
supported solid, we can refine the mesh of the solid given in the previous example.
We will subdivide each one of the two elements into 8 elements. To do so, add these
steps after step 7, then continue with step 8 above.

   8. Under ⎢ Geometry & Mesh ⎥: **Operations** select Subdivide.

   9. Set the first three rows to 2; 2; 2.

   10. Press Elements, then in ⑥ select the two elements with a (LC), then (RC).
Press OK.

11. Under ⎢ Geometry & Mesh ⎥: **Operations** select Sweep.

   12. Set Sweep/Tolerance to 0.01.

   13. Under Remove Unused, press Nodes. Press OK.

Results

Vertical translation of the load application points: $u_y = -0.0105237$.

Chapter 8
Elasto-Plastic Simulation

8.1 Fundamentals of Plastic Material Behavior

The following paragraph summarizes some fundamental characteristics and modelling approaches of one-dimensional elasto-plastic material behavior. The interested reader is referred to [3, 15, 16] for further information.

The characteristic feature of plastic material behavior is that a remaining strain ε^{pl} occurs after complete unloading, see Fig. 8.1a.

Only the elastic strain ε^{el} returns to zero at complete unloading. An additive composition of the strains by their elastic and plastic parts

$$\varepsilon = \varepsilon^{el} + \varepsilon^{pl} \tag{8.1}$$

is permitted at restrictions to small strains. The elastic strain ε^{el} can hereby be determined via HOOKE's law, i.e. $\sigma = E\varepsilon^{pl}$. The constitutive description of plastic material behavior includes

- a yield condition,
- a flow rule and
- a hardening law.

The yield condition enables to determine whether the relevant material suffers only elastic or also plastic strains at a certain stress state of the relevant body. In the uniaxial tensile test, plastic flow begins when reaching the initial yield stress k^{init}, see Fig. 8.1. A simplified form of the one-dimensional yield condition can be written as

$$F(\sigma, \kappa) = |\sigma| - k(\kappa) \leq 0, \tag{8.2}$$

where $k(\kappa)$ is an experimental material parameter, the so-called flow stress, and κ is the inner variable of isotropic hardening. In the case of ideal plasticity, see Fig. 8.1b, the following is valid: $F = F(\sigma)$. The flow rule serves as a mathematical description of the evolution of the infinitesimal increments of the plastic strain $d\varepsilon^{pl}$ in the course

© Springer International Publishing AG 2018
A. Öchsner and M. Öchsner, *A First Introduction to the Finite Element Analysis Program MSC Marc/Mentat*, https://doi.org/10.1007/978-3-319-71915-3_8

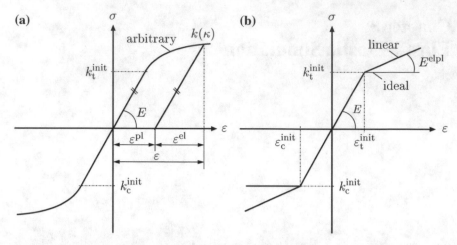

Fig. 8.1 Uniaxial stress-strain diagrams for different isotropic hardening laws: **a** arbitrary hardening; **b** linear hardening and ideal plasticity

of the load history of the body. The simplest form is the so-called associated flow rule

$$d\varepsilon^{pl} = d\lambda \frac{\partial F(\sigma, \kappa)}{\partial \sigma} = d\lambda \, \text{sgn}(\sigma) \,, \tag{8.3}$$

where the factor $d\lambda$ is the consistency parameter ($d\lambda \geq 0$) and $\text{sgn}(\sigma)$ represents the so-called sign function. The hardening law allows the consideration of the influence of material hardening on the yield condition and the flow rule. In the case of isotropic hardening, the yield stress is expressed as being dependent on an inner variable κ:

$$k = k(\kappa) \,. \tag{8.4}$$

If the equivalent plastic strain is used for the hardening variable ($\kappa = |\varepsilon^{pl}|$), then one talks about strain hardening. Another possibility is to describe the hardening being dependent on the specific plastic work ($\kappa = w^{pl} = \int \sigma d\varepsilon^{pl}$). Then one talks about work hardening. If Eq. (8.4) is combined with the flow rule according to (8.3), the evolution equation for the isotropic hardening variable results in:

$$d\kappa = d|\varepsilon^{pl}| = d\lambda \,. \tag{8.5}$$

Figure 8.2 shows the flow curve, i.e. the graphical illustration of the yield stress being dependent on the inner variable for different hardening approaches. This formulation must be provided in MSC Marc.

Fig. 8.2 Flow curve for different isotropic hardening laws

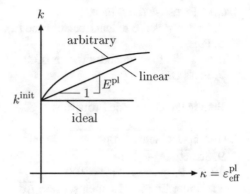

8.2 Basic Examples

8.2.1 Tensile Sample with Ideal-Plastic Material Behavior

Consider a cantilever bar which is discretized with a single linear bar finite element (use MSC Marc element type 9) as shown in Fig. 8.3. The material behavior is elastic (characterized by YOUNGS's modulus E) and ideal-plastic (characterized by the tensile yield stress, see Fig. 8.3a). A linearly increasing displacement (final value: $u = 8 \times 10^{-3}$ m) is applied at the right-hand end in 10 equal increments, see Fig. 8.3b. Calculate for each increment the stress, total strain, plastic strain, and the reaction force.

Marc solution

Under File → Save As..., save file as 'bar_id_plast'.
  Constructing the mesh

1. Under $\boxed{\text{Geometry \& Mesh}}$: **Basic Manipulation** select Geometry and Mesh.
  2. Under Mesh\Nodes, select $\overline{\text{Add}}$.

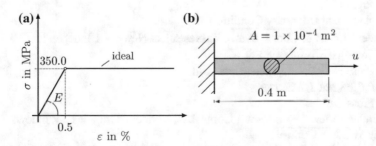

Fig. 8.3 Tensile sample with ideal-plastic material behavior: **a** stress-strain diagram; **b** geometry and boundary conditions

3. *0,0,0* ENTER *400,0,0* ENTER.
    4. Under Mesh\Elements, select $\overline{\text{Line (2)}}$.
    5. Press $\overline{\text{Add}}$.
6. In ⑥, select the two nodes with a (LC).
    7. Press $\overline{\text{OK}}$.

Setting the Geometric Properties

8. Under $\boxed{\text{Geometric Properties}}$: **Geometric Properties** select New (Structural).
    9. Select 3D → Truss.
    10. Set Properties\Area = 100.
    11. Under Entities\Elements, press $\overline{\text{Add}}$.
12. In ⑥ select the element with a (LC), then (RC).
    13. Press $\overline{\text{OK}}$.

Setting the Material Properties

14. Under $\boxed{\text{Material Properties}}$: **Material Properties** select New → Finite Stiffness Region → Standard.
    15. Set Other Properties\Young's Modulus = 70000 (see Fig. 8.4).
    16. Other Properties Press $\overline{\text{Plasticity}}$. Tick Plasticity (see Fig. 8.4). Set Yield Stress = 350. Press $\overline{\text{OK}}$ (see Fig. 8.5).

    17. Under Entities\Elements, press $\overline{\text{Add}}$
18. In ⑥ select the element with a (LC), then (RC).
    19. Press $\overline{\text{OK}}$

Setting the Boundary Conditions

Fixed Support
20. Under $\boxed{\text{Boundary Conditions}}$: **Boundary Conditions** select New (Structural) → Fixed Displacement.
    21. Under Properties tick Displacement X, Displacement Y and Displacement Z.
    22. Under Entities\Nodes, press $\overline{\text{Add}}$.
23. Under ⑥ select the left most node *(0,0,0)* with a (LC), then (RC).
    24. Press $\overline{\text{OK}}$.
    Displacement Boundary Condition
25. Under $\boxed{\text{Tables \& Coord. Syst.}}$: **Tables** select New → 1 Independent Variable.
    26. Set Type = time (see Fig. 8.6).
    27. Press $\overline{\text{Data Points}}$, then select $\overline{\text{Add}}$.
    28. *0,0* ENTER *1,1* ENTER.
    29. Press $\overline{\text{OK}}$.
30. Under $\overline{\text{View}}$ → View Control..., under Graphics Windows select Graphic Window Control.
    31. Under Show/Hide/Activate Windows, untick Table. Press $\overline{\text{OK}}$.

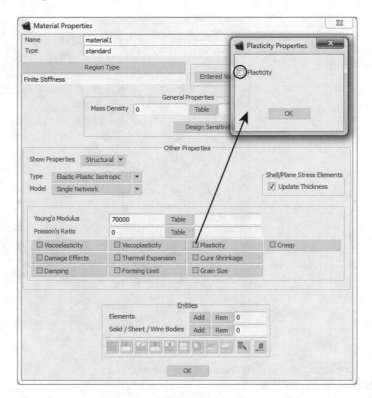

Fig. 8.4 Definition of elasto-plastic material properties

32. Press $\overline{OK}$.
33. Under ⎹Boundary Conditions⎸: **Boundary Conditions** select New (Structural)
→ Fixed Displacement. Set Name = disp.
   34. Under Properties tick Displacement X. Set Displacement X = 8.
   35. Under Properties, press $\overline{Table}$. Select table (see Fig. 8.7).
   36. Under Entities\Nodes, press $\overline{Add}$.
37. Under ⑥ select the right most node *(400,0,0)* with a (LC), then (RC).
   38. Press $\overline{OK}$.

Running the Job

39. Under ⎹Loadcases⎸: **Loadcases** select New → Static.
   40. Set Name = fixed_displacement.
   41. Set Stepping Procedure\#Steps = 10. Press $\overline{OK}$.

42. Under ⎹Jobs⎸: **Jobs** select New → Structural.
   43. Under Available, select fixed_displacement.
   44. Press $\overline{Job\ Results}$. Under Available Element Tensors, tick Stress, Total Strain, and Plastic Strain. Then press $\overline{OK}$.

Fig. 8.5 Definition of
plastic material properties:
yield criterion, hardening
rule, and yield stress

45. Press Check; See in ⑧ if there are any errors.
46. If there are none, press Run.
   47. In "Run Job", press Advanced Job Submission.
      48. Press Save Model.
      49. Press Write Input File. Press OK.
      50. Press Submit 1.
   51. Wait until Status = Complete.
   52. Press Open Post File (Model Plot Results Menu).

Viewing the model

53. Under Deformed Shape\Style, select Deformed and Original
54. Under Scalar Plot\Style, select Numerics.
55. Under Scalar Plot, press Scalar and select Displacement X.
56. Use in ⑦ the buttons for 'play' (▶), 'next increment' (▶|), 'previous increment' (|◀) etc. to select different increments.
57. Under Scalar Plot, press Scalar and successively select Comp 11 of Stress, Comp 11 of Total Strain, Comp 11 of Plastic Strain, and Reaction Force X for the evaluation.
58. Press OK.

Fig. 8.6 Defining a linear function based on table input

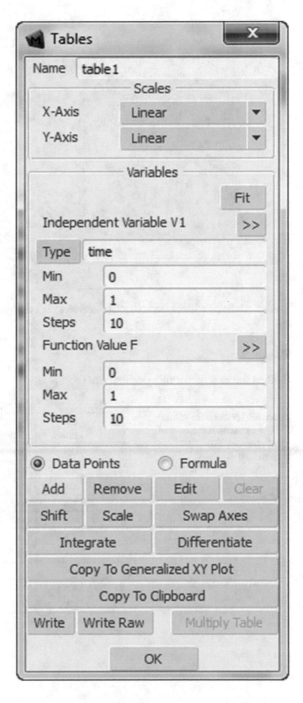

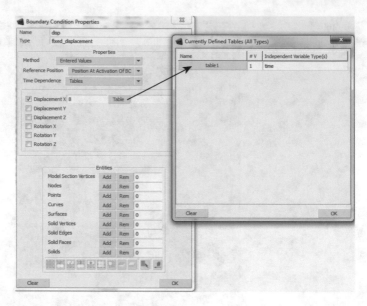

Fig. 8.7 Defining a linear increasing displacement via a table

Result:

Collect the results for this problem in Table 8.1.

Table 8.1 Numerical values for one element in the case of ideal plasticity (10 increments; $\Delta u_2 = 0.8$ mm)

Inc	u_2	σ	ε	ε^{pl}	F_R
–	mm	MPa	10^{-2}	10^{-2}	kN
1	0.8	140	0.2	0.0	14
2	x.x	xxx	x.x	x.x	xx
3	x.x	xxx	x.x	x.x	xx
4	x.x	xxx	x.x	x.x	xx
5	x.x	xxx	x.x	x.x	xx
6	x.x	xxx	x.x	x.x	xx
7	x.x	xxx	x.x	x.x	xx
8	x.x	xxx	x.x	x.x	xx
9	x.x	xxx	x.x	x.x	xx
10	x.x	xxx	x.x	x.x	x

8.2.2 Tensile Sample with Linear Hardening

Consider a cantilever bar which is discretized with a single linear bar finite element (use MSC Marc element type 9) as shown in Fig. 8.8. The material behavior is elastic ($E = 70000$ MPa) and plastic with isotropic linear hardening (characterized by the plastic modulus E^{pl}, see Fig. 8.8a). The load on the right-hand end of the bar is applied in 10 equal increments as (a) a displacement $u = 8 \times 10^{-3}$ m or (b) as a point load $F = 100$ kN.

Calculate for each increment the stress, the total strain, the plastic plastic strain, and the reaction force.

Marc solution

The solution steps are very similar to example in Sect. 8.2.1. The major difference is that the hardening behavior must be defined. Thus, the following modification must be considered after step 13:

Setting the Material Properties

14. Under $\boxed{\text{Tables \& Coord. Syst.}}$: **Tables** select → 1 Independent Variable.
    15. Set Name = plast.
    16. Set Type = eq_plastic_strain (see Fig. 8.9).
    17. Under Independent Variable V1: Set Min = 0 and set Max = 0.1.
    18. Under Function Value F: Set Min = 0 and set Max = 700.
    19. Press Data Points, then select Add.
    20. *0,350* ENTER *0.04091,636.3636* ENTER.
    21. Press $\overline{\text{OK}}$.

22. Under View → View Control..., under Graphics Windows select Graphic Window Control.
    23. Under Show/Hide/Activate Windows, untick Table. Press $\overline{\text{OK}}$.

24. Under $\boxed{\text{Material Properties}}$: **Material Properties** select New → Finite Stiffness Region → Standard.

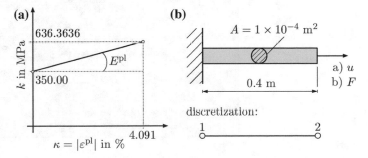

Fig. 8.8 Tensile sample with linear hardening: **(a)** flow curve; **(b)** geometry and boundary conditions

Fig. 8.9 Defining a linear
function for the flow curve
definition (see Fig. 8.3a)
based on table input

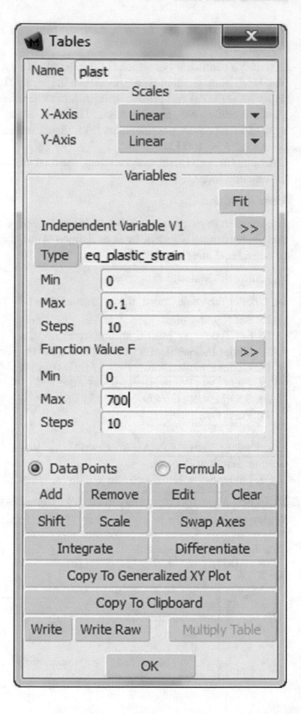

25. Set Other Properties\Young's Modulus = 70000.
26. Other Properties Press $\overline{\text{Plasticity}}$. Tick Plasticity. Set Yield Stress = 1.
    Press $\overline{\text{Table}}$ and select table plast (see Fig. 8.10). Press $\overline{\text{OK}}$.
27. Press $\overline{\text{OK}}$.

Result

Collect the results for this problem in Tables 8.2 and 8.3.

Fig. 8.10 Defining a linear flow curve (see Fig. 8.3a)

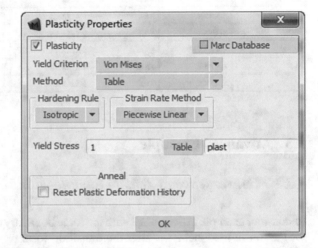

Table 8.2 Numerical values for one element in the case of linear hardening (10 increments; $\Delta u_2 = 0.8$ mm)

Inc –	u_2 mm	σ MPa	ε 10^{-2}	ε^{pl} 10^{-2}	F_{R} kN
1	0.8	140	0.2	0.0	14
2	x.x	xxx	x.x	x.x	xx
3	x.x	xxx	x.x	x.x	xx
4	x.x	xxx	x.x	x.x	xx
5	x.x	xxx	x.x	x.x	xx
6	x.x	xxx	x.x	x.x	xx
7	x.x	xxx	x.x	x.x	xx
8	x.x	xxx	x.x	x.x	xx
9	x.x	xxx	x.x	x.x	xx
10	x.x	xxx	x.x	x.x	xx

Table 8.3 Numerical values for one element in the case of linear hardening (10 increments; $\Delta F_2 = 1 \times 10^4$ N)

Inc	u_2	σ	ε	ε^{pl}	F_R
–	mm	MPa	10^{-2}	10^{-2}	kN
1	0.5714	100	0.1429	0.0	0
2	X.XXXX	XXX	X.XXXX	X.X	X
3	X.XXXX	XXX	X.XXXX	X.X	X
4	X.XXXX	XXX	X.XXXX	X.X	X
5	X.XXXX	XXX	X.XXXX	X.X	X
6	X.XXXX	XXX	X.XXXX	X.X	X
7	X.XXXX	XXX	X.XXXX	X.X	X
8	X.XXXX	XXX	X.XXXX	X.X	X
9	X.XXXX	XXX	X.XXXX	X.X	X
10	X.XXXX	XXX	X.XXXX	X.X	X

8.3　Advanced Example

8.3.1　Convergence of Tensile Sample with Linear Hardening

Reconsider example in Sect. 8.2.2 with the force boundary condition. The stress value for the fourth increment, i.e. after the transition form the pure elastic to the elasto-plastic regime, is obtained as $\sigma = 372.284$ MPa. However, the stress value based on the external load for the fourth increment ($F = 4 \times 10^4$ N) would result in a stress of

$$\sigma = \frac{F}{A} = \frac{4 \times 10^4 \text{ N}}{100 \text{ mm}^2} = 400 \text{ MPa}. \tag{8.6}$$

Change the convergence testing from the default setting to a maximum absolute displacement value of 1×10^{-5} mm in order to improve the accuracy of the iterative procedure.

Marc solution

The solution steps a very similar to the example in Sect. 8.2.2. The major difference is that that convergence testing must be re-defined. Thus, the following modification must be considered for step 39 (numbering according to the example in Sect. 8.2.1):

39. Under ⎿Loadcases⏌: **Loadcases** select New → Static.
   40. Set Name = lin_force.
   41. Press Convergence Testing.
      42. Tick Absolute and Displacements (see Fig. 8.11).
      43. Under Displacements, set Maximum Absolute Displacement = 0.00001.
      44. Press OK.
   45. Set Stepping Procedure\#Steps = 10. Press OK.

Fig. 8.11 Updating the
parameters for convergence
testing

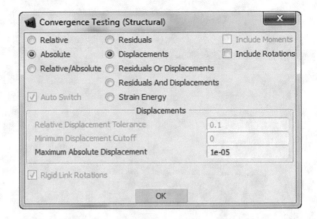

Result

The modified convergence testing parameters result now in a stress of $\sigma = 400$ for
increment 4. This is now equal to the theoretical value given in Eq. (8.6).

Chapter 9
Advanced Topics

9.1 Procedure Files

Procedure files are text files (ASCII) which allow you to execute a pre-defined sequence of commands[1] (see the example in Fig. 9.1). They can be useful, for example, to generate a mesh without having to input nodes and elements manually. Whenever writing a procedure file in a text editor, they have to be saved as a '.proc' file.

The following commands are of use when writing a procedure file:
| a vertical bar introduces a comment, i.e. a string of text which is not compiled.
*, a star is used to denote commands.
#, the hash symbol ends lists.

Executing a procedure file

After saving a procedure file, there are two ways to execute it.

The first option is to go to Tools → Procedures... in Marc (see Fig. 9.2). Press $\overline{\text{Load}}$ and select your procedure file. Press $\overline{\text{Start/Cont}}$ (see Fig. 9.3). Wait for Marc to finish executing the program. Then, press $\overline{\text{OK}}$.

The other option is to simply click on your procedure file. In some cases it might be necessary to specify the program to run it with as being 'mentat', or 'mentat.bat', as it may appear. Marc will be opened and the procedure file should start to run.

---

[1]The exact syntax of Marc commands is shown in the Command Line Dialog when using them, cf. Fig. 1.4, ⑧.

© Springer International Publishing AG 2018
A. Öchsner and M. Öchsner, *A First Introduction to the Finite Element Analysis
Program MSC Marc/Mentat*, https://doi.org/10.1007/978-3-319-71915-3_9

```
transmission_tower.proc - Notepad
File  Edit  Format  View  Help
|| Created by Marc Mentat 2013.1.0 (b) (64bit)
*prog_option compatibility:prog_version:ment2013.1
*prog_analysis_class structural

*add_nodes
6.000,0.000,0
2.000,12.00,0
-1.00,12.00,0
4.000,12.00,0
7.000,12.00,0
0.700,12.00,0
5.300,12.00,0
2.000,13.00,0
4.000,13.00,0
2.000,14.00,0
4.000,14.00,0
-1.00,14.00,0
7.000,14.00,0
```

Fig. 9.1 Example of a procedure file

Fig. 9.2 Accessing the
procedure file window

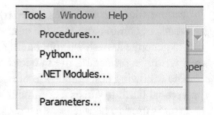

9.2   Command-Line Execution

The solver MSC Marc can be independently operated from the graphical interface
MSC Mentat. This may result in a much better use of the existing hardware (RAM)
because the model in not opened in the pre-processor and much more memory can
be allocated to the equation solver. The data input file for the solver (*.dat) can
be submitted from a command prompt window[2] (see Fig. 9.4) to the solver via a
specific command. However, the user must first navigate to the working directory,
i.e. the directory where the model and all files are saved.

Table 9.1 summarizes a few basic commands for command line operations.

To actually submit a Marc job, the following command should be used:

$$\texttt{run\_marc -jid jobidname}$$

---

[2]A command prompt window can be opened by clicking Start → All Programs → Accessories →
Command Prompt.

Fig. 9.3 Procedure file window

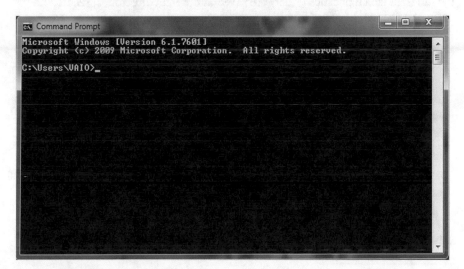

Fig. 9.4 Command prompt window

Table 9.1 Useful commands for command prompt operations

Command	Explanation
dir	Lists the files and directories in the folder that you are in
cd	Changes directory relevant to the one you are currently working in
cd..	Just goes back one level

This runs the job 'jidname.dat' in the current working directory. Depending on the computer installation, it might be required to set the following value for the `path` variable[3]:

```
C:\MSC.Software\Marc\2014.0.0\marc2014\tools;
C:\MSC.Software\Marc\2014.0.0\marc2014\bin;
```

9.3 Batch Processing

The information of the previous section can be used to run the solver in batch processing mode, i.e. the execution of a series of runs ('jobs') without further user intervention. This is quite useful in the case of parametric studies. Let us assume that the different data input files (*.dat) are saved in different directories, i.e. 01, 02, 03. Then, the following batch file[4]

```
CLS
    cd 01
    start /low /b /wait run_marc -jid jidname_01
    cd ..
    cd 02
    start /low /b /wait run_marc -jid jidname_02
    cd ..
    cd 03
    start /low /b /wait run_marc -jid jidname_03
    cd ..
pause
```

solves sequential three models without further interaction.

---

[3]The environment variables window can be accessed by clicking Start → Control Panel → System → Advanced system settings → System Properties → Advanced → Environment Variables ... → System variables → Variable Path.

[4]Simply save an ASCII file with the extension *.bat.

Appendix A
Answers to Additional Questions

Questions from Chap. 2

2.2.1 1D Rod—Fixed Displacement

• One-element solutions:

Marc: $\varepsilon = 0.5$, $\sigma = 10$.

Comment: Stress and strain is not a default output and must be selected:

Under $\boxed{\text{Jobs}}$: **Jobs** select New $\rightarrow$ Structural.

In "Run Job", press $\overline{\text{Job Results}}$ and select the required Available Element Tensors ('Stress' and 'Total Strain').

Analytical solution: $\varepsilon = \frac{\Delta L}{L} = \frac{0.5}{1.0} = 0.5$, $\sigma = E\varepsilon = 20 \times 0.5 = 10$.

• Two-element solutions:

Marc: $u_{x2} = 0.25$, $u_{x3} = 0.5$, $\sigma_{x2} = \sigma_{x3} = 10$, $\varepsilon_{x2} = \varepsilon_{x3} = 0.5$.

FE hand calculation: $u_{2x} = \frac{1}{2}u_0$, $u_{3x} = u_0$.

2.2.2 1D Rod—Fixed Point Load

• One-element solutions:

Marc: $\varepsilon = 0.5$, $\sigma = 10$.

Analytical solution: $\sigma = \frac{F_0}{A} = \frac{5}{0.5} = 10$, $\sigma = E\varepsilon \rightarrow \varepsilon = \frac{\sigma}{E} = \frac{10}{20} = 0.5$.

• Two-element solutions:

Marc: $u_{x2} = 0.25$, $u_{x3} = 0.5$, $\sigma_{x2} = \sigma_{x3} = 10$, $\varepsilon_{x2} = \varepsilon_{x3} = 0.5$.

FE hand calculation: $u_{2x} = \frac{F_0 L'}{EA}$, $u_{3x} = \frac{2F_0 L'}{EA}$.

2.2.4 Plane Truss—Triangle

• Stress in each rod element:

$$\sigma_{1x} = \frac{E}{L}(u_{2x} - u_{1x}), \tag{A.1}$$

$$= ([\cos\alpha u_{2X} + \sin\alpha u_{2Y}] - [\cos\alpha u_{1X} + \sin\alpha u_{1Y}]). \tag{A.2}$$

© Springer International Publishing AG 2018
A. Öchsner and M. Öchsner, *A First Introduction to the Finite Element Analysis Program MSC Marc/Mentat*, https://doi.org/10.1007/978-3-319-71915-3

Table A.1 Results for F

Element	Angle α	u_{endX}	u_{endY}	u_{startX}	u_{startY}	Elemental stress σ
I	30°	0.0166668	−0.00577367	0	0	0.2309
II	90°	0.0166668	−0.00577367	0	0	−0.1155
III	330°	0	0	0	0	0

Table A.2 Results for u

Element	Angle α	u_{endX}	u_{endY}	u_{startX}	u_{startY}	Elemental stress σ
I	30°	0.01	−0.00346418	0	0	0.1386
II	90°	0.01	−0.00346418	0	0	−0.06928
III	330°	0	0	0	0	0

The results for the force and displacement boundary conditions are summarized in Tables A.1 and A.2.

• Procedure file:

```
*add_nodes
-0.866,0.500,0
0,1.000,0
0,0,0

*fill_view
*set_element_class line2
*add_elements
 1
 2
 2
 3
 1
 3
```

• finite element 'hand calculation'

Force boundary condition:

$$
\begin{bmatrix} u_{2X} \\ u_{2Y} \\ u_{3X} \end{bmatrix} = \frac{L}{EA} \begin{bmatrix} \frac{5}{3} & -\frac{\sqrt{3}}{3} & 0 \\ -\frac{\sqrt{3}}{3} & 1 & 0 \\ 0 & 0 & \frac{4}{3} \end{bmatrix} \begin{bmatrix} F \\ 0 \\ 0 \end{bmatrix} = \frac{LF}{EA} \begin{bmatrix} \frac{5}{3} \\ -\frac{\sqrt{3}}{3} \\ 0 \end{bmatrix}. \tag{A.3}
$$

The rod stresses can be obtained from the global displacements as:

$$\sigma_{\mathrm{I}} = \frac{F_{\mathrm{I}}}{A} = \frac{E}{L}(-\cos\alpha_{\mathrm{I}}u_{1X} - \sin\alpha_{\mathrm{I}}u_{1Y} + \cos\alpha_{\mathrm{I}}u_{2X} + \sin\alpha_{\mathrm{I}}u_{2Y}) = \frac{2\sqrt{3}\,F}{3A}, \quad (A.4)$$

$$\sigma_{\mathrm{II}} = \frac{F_{\mathrm{II}}}{A} = \frac{E}{L}(-\cos\alpha_{\mathrm{II}}u_{3X} - \sin\alpha_{\mathrm{II}}u_{3Y} + \cos\alpha_{\mathrm{II}}u_{2X} + \sin\alpha_{\mathrm{II}}u_{2Y}) = \frac{\sqrt{3}\,F}{3A}, \quad (A.5)$$

$$\sigma_{\mathrm{III}} = \frac{F_{\mathrm{III}}}{A} = \frac{E}{L}(-\cos\alpha_{\mathrm{III}}u_{1X} - \sin\alpha_{\mathrm{III}}u_{1Y} + \cos\alpha_{\mathrm{III}}u_{3X} + \sin\alpha_{\mathrm{III}}u_{3Y}) = 0. \quad (A.6)$$

Displacement boundary condition:

$$\begin{bmatrix} u_{2X} \\ u_{2Y} \\ u_{3X} \end{bmatrix} = \frac{L}{EA} \begin{bmatrix} \frac{EA}{L} & 0 & 0 \\ -\frac{\sqrt{3}EA}{5L} & \frac{4}{5} & 0 \\ 0 & 0 & \frac{4}{3} \end{bmatrix} \begin{bmatrix} u \\ 0 \\ 0 \end{bmatrix} = u \begin{bmatrix} 1 \\ -\frac{\sqrt{3}}{5} \\ 0 \end{bmatrix}. \quad (A.7)$$

Reactions and rod stresses can be obtained as:

$$\left(R_{1X} = -\frac{3}{5} \times \frac{EAu}{L}\right), \quad \left(R_{1Y} = -\frac{\sqrt{3}}{5} \times \frac{EAu}{L}\right), \quad R_{2X} = \frac{3}{5} \times \frac{EAu}{L}, \quad (A.8)$$

$$\left(R_{3X} = 0, \quad R_{3Y} = \frac{\sqrt{3}}{5} \times \frac{EAu}{L}\right). \quad (A.9)$$

$$\sigma_{\mathrm{I}} = \frac{F_{\mathrm{I}}}{A} = \frac{2\sqrt{3}}{5} \times \frac{Eu}{L}, \quad \sigma_{\mathrm{II}} = \frac{F_{\mathrm{II}}}{A} = \frac{\sqrt{3}}{5} \times \frac{Eu}{L}, \quad \sigma_{\mathrm{III}} = \frac{F_{\mathrm{III}}}{A} = 0. \quad (A.10)$$

2.3.2 Transmission Tower Structure

$$F_X = \pm 39.4227, \; F_Y = 300. \quad (A.11)$$

2.4.1 Truss Structure with Six Members

The following equations allow to calculate the numerical values for the general case and the two special cases ($u_0 = 0 \vee F_0 = 0$).

- Displacements of the nodes:

$$u_{2X} = 0.429\,u_0 - 0.408\,\frac{FL}{EA}, \quad (A.12)$$

$$u_{2Z} = -1.357\,u_0 - 1.562\,\frac{FL}{EA}, \tag{A.13}$$

$$u_{4X} = 0.285\,u_0 - 0.296\,\frac{FL}{EA}, \tag{A.14}$$

$$u_{4Z} = -0.928\,u_0 - 1.970\,\frac{FL}{EA}. \tag{A.15}$$

• Reaction forces at the supports and nodes where displacements are prescribed:

$$R_{1X} = -0.184\,F_0 - 0.858\,\frac{EAu_0}{L}, \tag{A.16}$$

$$R_{1Z} = 0.592\,F_0 + 0.429\,\frac{EAu_0}{L}, \tag{A.17}$$

$$R_{3X} = 0.296\,F_0 - 0.285\,\frac{EAu_0}{L}, \tag{A.18}$$

$$R_{3Z} = 0, \tag{A.19}$$

$$R_{5X} = -0.112\,F_0 + 1.144\,\frac{EAu_0}{L}, \tag{A.20}$$

$$R_{5Z} = -0.408\,F_0 + 0.429\,\frac{EAu_0}{L}. \tag{A.21}$$

• Stress and strain in each element:

$$\sigma_{\mathrm{I}} = -0.408\,\frac{F_0}{A} + 0.429\,\frac{Eu_0}{L}, \tag{A.22}$$

$$\sigma_{\mathrm{II}} = -0.296\,\frac{F_0}{A} + 0.285\,\frac{Eu_0}{L}, \tag{A.23}$$

$$\sigma_{\mathrm{III}} = 1.184\,\frac{F_0}{A} + 0.858\,\frac{Eu_0}{L}, \tag{A.24}$$

$$\sigma_{\mathrm{IV}} = 0.408\,\frac{F_0}{A} - 0.429\,\frac{Eu_0}{L}, \tag{A.25}$$

$$\sigma_{\mathrm{V}} = 0.296\,\frac{F_0}{A} + 0.715\,\frac{Eu_0}{L}, \tag{A.26}$$

$$\sigma_{\mathrm{VI}} = -0.816\,\frac{F_0}{A} + 0.858\,\frac{Eu_0}{L}. \tag{A.27}$$

$$\varepsilon_{\mathrm{I}} = -0.408 \frac{F_0}{EA} + 0.429 \frac{u_0}{L}, \tag{A.28}$$

$$\varepsilon_{\mathrm{II}} = -0.296 \frac{F_0}{EA} + 0.285 \frac{u_0}{L}, \tag{A.29}$$

$$\varepsilon_{\mathrm{III}} = 1.184 \frac{F_0}{EA} + 0.858 \frac{u_0}{L}, \tag{A.30}$$

$$\varepsilon_{\mathrm{IV}} = 0.408 \frac{F_0}{EA} - 0.429 \frac{u_0}{L}, \tag{A.31}$$

$$\varepsilon_{\mathrm{V}} = 0.296 \frac{F_0}{EA} + 0.715 \frac{u_0}{L}, \tag{A.32}$$

$$\varepsilon_{\mathrm{VI}} = -0.816 \frac{F_0}{EA} + 0.858 \frac{u_0}{L}. \tag{A.33}$$

Questions from Chap. 3

3.2.1 Beam with a Square Cross-Section

- Force (F_y) boundary condition:

$$u_y(x) = \frac{F_y}{EI_z}\left(\frac{1}{6}x^3 - \frac{1}{2}Lx^2\right), \tag{A.34}$$

$$\varphi_z(x) = \frac{F_y}{EI_z}\left(\frac{1}{2}x^2 - Lx\right), \tag{A.35}$$

$$M_z(x) = F_y(x - L), \tag{A.36}$$

$$Q_y(x) = -F_y. \tag{A.37}$$

Values at node 2: $u_y(L) = -\frac{F_y L^3}{3EI_z}$, $\varphi_z(L) = -\frac{F_y L^2}{2EI_z}$, $M_z(L) = 0$, $Q_y(L) = -F_y$.

- Displacement (u_y) boundary condition:

$$u_y(x) = u_y\left(\frac{1}{2}\left(\frac{x}{L}\right)^3 - \frac{3}{2}\left(\frac{x}{L}\right)^2\right), \tag{A.38}$$

$$\varphi_z(x) = u_y\left(\frac{3}{2}\frac{x^2}{L^3} - 3\frac{x}{L^2}\right), \tag{A.39}$$

$$M_z(x) = \frac{3EI_z u_y}{L^2}\left(\frac{x}{L} - 1\right), \tag{A.40}$$

$$Q_y(x) = -\frac{3EI_z u_y}{L^3}. \tag{A.41}$$

Values at node 2: $u_y(L) = -u_y$, $\varphi_z(L) = -\frac{3u_y}{2L}$, $M_z(L) = 0$, $Q_y(L) = -\frac{3EI_z u_y}{L^3}$.

• Moment (M_z) boundary condition:

$$u_y(x) = -\frac{M_z x^2}{2EI_z},$$ (A.42)

$$\varphi_z(x) = -\frac{M_z x}{EI_z},$$ (A.43)

$$M_z(x) = M_z,$$ (A.44)

$$Q_y(x) = 0.$$ (A.45)

Values at node 2: $u_y(L) = -\frac{M_z L^2}{2EI_z}$, $\varphi_z(L) = -\frac{M_z L}{EI_z}$, $M_z(L) = M_z$, $Q_y(L) = 0$.

• Rotation (φ_z) boundary condition:

$$u_y(x) = -\frac{\varphi_z x^2}{2L},$$ (A.46)

$$\varphi_z(x) = -\frac{\varphi_z x}{L},$$ (A.47)

$$M_z(x) = -\frac{\varphi_z EI_z}{L},$$ (A.48)

$$Q_y(x) = 0.$$ (A.49)

Values at node 2: $u_y(L) = -\frac{\varphi_z L}{2}$, $\varphi_z(L) = -\varphi_z$, $M_z(L) = -\frac{\varphi_z EI_z}{L}$, $Q_y(L) = 0$.

3.3.1 Plane Bridge Structure with Beam Elements

Deformation at node 3: $u_{3Y} = -0.965684$, $u_{3X} = 0$.

3.3.2 Transmission Tower with Beam Elements

The vertical displacement is as follows:

upper nodes	−0.00109
middle nodes	−0.00104
lower nodes	−0.00095

Reaction forces at each foundation node: $F_{RY} = 300$.

3.3.3 Beam Element—Generalized Strain and Stress Output

Analytical solution for bending in the xy-plane, x-axis along the principal beam axis:

• $u_y(x) = \frac{1}{EI_z}\left(\frac{F_y x^3}{6} - \frac{F_y L x^2}{2}\right)$,
• $M_z(x) = F_y(x - L)$,
• $Q_y(x) = -F_y$,
• $\kappa_z(x) = \frac{F_y}{EI_z}(x - L)$,

- $|\sigma_{\max}(0)| = \frac{F_y L a/2}{I_z}, |\sigma_{\max}(L)| = 0,$
- $|\varepsilon_{\max}(0)| = \frac{F_y L a/2}{E I_z}, |\varepsilon_{\max}(L)| = 0.$

Marc solution (Vector defining local zx-plane chosen as [0,1,0], i.e. local x-axis):

Quantity	Node 1	Node 2	Real Meaning
σ_{11}	0	0	–
σ_{22}	0	0	–
σ_{33}	10	10	N_X or N_z
σ_{12}	0	0	–
σ_{23}	−499.807	−0.192576	M_Z or M_y
σ_{31}	249.904	0.0962879	M_Y or M_x

Quantity	Node 1	Node 2	Real Meaning
ε_{11}	0	0	–
ε_{22}	0	0	–
ε_{33}	0.0002	0.0002	ε_X or ε_z
ε_{12}	0	0	–
ε_{23}	−0.479815	−0.000184873	κ_Z or κ_y
ε_{31}	0.239908	0.0000924	κ_Y or κ_z

Elemental values are evaluated at the integration points (3 for Marc element type 52) and extrapolated to the nodes (the above values are based on the 'Linear' method).

3.3.4 Beam Element—Generalized Stresses for Multiaxial Stress State

The different configurations and corresponding stress states are shown in Fig. A.1.

Based on analytical mechanics, the stresses result from the internal reactions as follows:

$$|\sigma_x| = \frac{M}{I} \times y, \quad \text{(bending)} \tag{A.50}$$

$$\sigma_x = \frac{N}{A}, \quad \text{(tension)} \tag{A.51}$$

$$\tau_{xz} = \frac{T}{I_p} \times r. \quad \text{(torsion)} \tag{A.52}$$

The different stress components, i.e. normal stress σ and shear stress τ, can be transformed to an effective stress σ_{eff}, which can be compared to the tensile yield stress k_t:

$$F_{\sigma-\tau} = \sigma_{\text{eff}} - k_t = \sqrt{\sigma^2 + 4\tau^2} - k_t = 0, \quad \text{(TRESCA)} \tag{A.53}$$

$$F_{\sigma-\tau} = \sigma_{\text{eff}} - k_t = \sqrt{\sigma^2 + 3\tau^2} - k_t = 0. \quad \text{(VON MISES)} \tag{A.54}$$

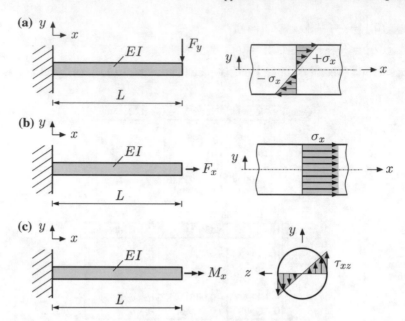

Fig. A.1 Cantilever beam loaded by different point loads and corresponding stress states: **a** bending in x-y plane, **b** tension in x-direction and **c** torsion around x-axis

Questions from Chap. 4

4.2.1 Timoshenko Beam with a Square Cross-Section

Analytical solutions (Table A.3):

$$u_{Y,\max}^{\mathrm{B}} = \frac{FL^3}{3EI}, \quad u_{Y,\max}^{\mathrm{T}} = \frac{FL^3}{3EI} + \frac{FL}{k_{\mathrm{s}}AG}. \tag{A.55}$$

4.3.1 Beam Element—Generalized Strain and Stress Output

Analytical solution:

$$u_y(x) = \frac{1}{EI_z}\left(-F\frac{x^3}{6} + FL\frac{x^2}{2} + \frac{EI_zF}{k_{\mathrm{s}}AG}x\right), \tag{A.56}$$

$$\phi_z(x) = \frac{1}{EI_z}\left(-F\frac{x^2}{2} + FLx\right), \tag{A.57}$$

Table A.3 Comparison of numerical and analytical results

	Case I, $L = 10 \times h$		Case II, $L = 0.2 \times h$	
	Euler-Bernoulli	Timoshenko	Euler-Bernoulli	Timoshenko
Analytical solution	-4	-4.0312	-0.000032	-0.000656
u_Y: 1 element	-4	-3.026	-0.000032	-0.000544
rel. error[a] in %	0%	24.936%	95.121%	17.0732%
u_Y: 5 elements	-4	-3.986	-0.000032	-0.00055168
rel. error in %	0%	1.121%	95.121%	15.902%
u_Y: 10 elements	-4	-4.016	-0.000032	-0.00055192
rel. error in %	0%	0.377%	95.121%	15.866%
u_Y: 50 elements	-4	-4.0196	-0.000032	-0.000551997
rel. error in %	0%	0.288%	95.121%	15.854%

[a]The relative error is calculated based on the formula $\left| \frac{\text{analytical solution - FE solution}}{\text{analytical solution}} \right| \times 100$. For the $L = 0.2 \times h$ case, the Timoshenko solution was taken as the analytical solution for both cases

$$M_z(x) = EI_z \frac{d\phi_z(x)}{dx}, \tag{A.58}$$

$$Q_y(x) = -\frac{dM_z(x)}{dx} = -EI_z \frac{d\phi_z^2(x)}{dx^2}. \tag{A.59}$$

Marc solution (Vector defining local zx-plane chosen as $[0,0,1]$, i.e. local x-axis):

Quantity	Node 1	Node 2	Real Meaning
σ_{11}	0	0	–
σ_{22}	0	0	–
σ_{33}	-100	-100	Q_Y or Q_y
σ_{12}	5	5	M_Z or M_x
σ_{23}	0	0	–
σ_{31}	0	0	–

Quantity	Node 1	Node 2	Real Meaning
ε_{11}	0	0	–
ε_{22}	0	0	–
ε_{33}	-0.00624	-0.00624	γ_{XY} or γ_{yz}
ε_{12}	0.0048	0.0048	κ_Z or κ_x
ε_{23}	0	0	–
ε_{31}	0	0	–

Elemental values are evaluated at the integration point (1 for Marc element type 98) and extrapolated to the nodes (the above values are based on the 'Linear' method).

Questions from Chap. 5

5.2.1 Plane Element Under Tensile Load

See Tables A.4 and A.5.

It must be noted here that the thickness strain ε_Z must be evaluated according to Eq. (5.1) since the software does not print this result.

5.2.2 Simply Supported Beam

See Table A.6.

Table A.4 Summary of numerical results for the *plane stress* problem

Quantity	Node 2	Node 3
u_X	0.0296517	0.0296517
u_Y	0.00283899	−0.00283899
σ_X	3881.85	3881.85
σ_Y	−358.555	−358.555
ε_Z	−0.0035233	−0.0035233
σ_Z	0	0

Table A.5 Summary of numerical results for the *plane strain* problem

Quantity	Node 2	Node 3
u_X	0.028723	0.028723
u_Y	0.00339267	−0.00339267
σ_X	3811.74	3811.74
σ_Y	−459.84	−459.84
ε_Z	0	0
σ_Z	670.38	670.38

Table A.6 Summary of numerical results for the simply supported beam problem

Case	Quantity	Node 2	Node 3
Top Loadcase	u_Y	−0.0435789	−0.0469123
	u_X	0.0180526	0.0180526
Bottom Loadcase	u_Y	−0.0517299	−0.0435789
	u_X	0.0170526	0.0170526
Split Loadcase	u_Y	−0.0476544	−0.0452456
	u_X	0.0175526	0.0175526

5.3.1 Mesh Refinement—Simply Supported Beam

See Table A.7.

5.3.2 Stress Concentration

The regular and biased meshed are shown in Figs. A.2 and A.3.

Regular Mesh	
Subdivision	Stress Concentration (σ_Y)
4×4	18.2835
8×8	23.431
16×16	28.3733
32×32	31.0081
Biased Mesh	
Subdivision	Stress Concentration (σ_Y)
4×4	25.9067
8×8	30.0995
16×16	31.267
32×32	31.2686

The analytical solution is obtained with $\sigma_{\mathrm{nom}} = \sigma \frac{A}{A-a}$ as: $\sigma_{\max} = 30.2349$.

5.3.3 Stress Intensity/Singularity   The regular and biased meshed are shown in

Regular Mesh	
Subdivision	Equivalent Stress (σ_{Mises})
4×4	571.62
8×8	784.35
16×16	1062.83
32×32	1447.35
Biased Mesh	
Subdivision	Equivalent Stress (σ_{Mises})
4×4	330.40
8×8	1083.76
16×16	1454.78
32×32	1980.23

Figs. A.4 and A.5.

5.3.4 Short Fiber Reinforced Composite Plate

First Step:

$$E_{\mathrm{m}} = \frac{\sigma_{\mathrm{m}}}{\varepsilon_{\mathrm{m}}} = \frac{\frac{300+300}{2 \times 5}}{\frac{0.1}{5}} = 3000, \tag{A.60}$$

$$E_{\mathrm{f}} = \frac{\sigma_{\mathrm{f}}}{\varepsilon_{\mathrm{f}}} = \frac{\frac{80}{0.1}}{\frac{0.1}{5}} = 40000. \tag{A.61}$$

Table A.7 Summary of numerical results for the mesh refinement problem

	2 Elements	8 Elements	32 Elements
Analytical Bernoulli	0.0675		
Analytical Timoshenko	0.0891		
Top Loadcase	0.0452456	0.0745145	0.09393464
Bottom Loadcase	0.0476544	0.0743642	0.09391202
Split Loadcase	0.04645	0.0740056	0.0942914

Fig. A.2 Regular meshes

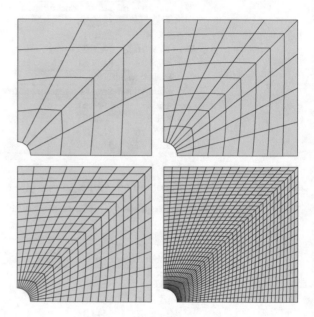

Second Step:

$$E = \frac{\sigma}{\varepsilon} = \frac{\frac{300+606+300}{10\times2+0.1}}{\frac{0.1}{5}} = 3000. \quad \text{(same material)} \tag{A.62}$$

$$E_c = \frac{\frac{300+680+300}{20.1}}{\frac{0.1}{5}} = \frac{6400}{2.01} = 3184.08. \quad \text{(different materials)} \tag{A.63}$$

The fiber volume fraction can be obtained based on the outer dimensions of the composite, i.e. $10 \times 5 \times 2$:

$$\phi_f = \frac{V_f}{V_c} = \frac{0.5}{100} = 0.005 = 0.5\%. \tag{A.64}$$

Fig. A.3 Biased meshes

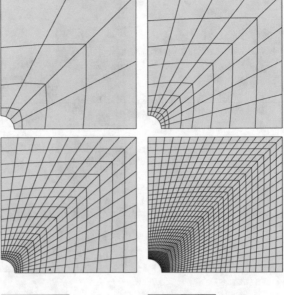

Fig. A.4 Regular meshes

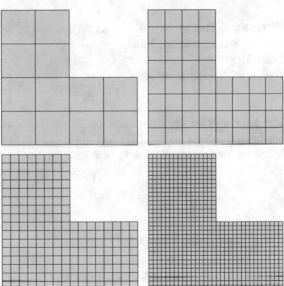

Thus, the upper and lower bounds according to the rule of mixture reads:

$$E_c = 3185.00 \text{ MPa}, \quad E_c = 3013.94 \text{ MPa}. \tag{A.65}$$

It should be noted here that the fiber volume fraction can be also obtained based on the composite volume which is assumed as the volume of the matrix plus the volume

Fig. A.5 Biased meshes

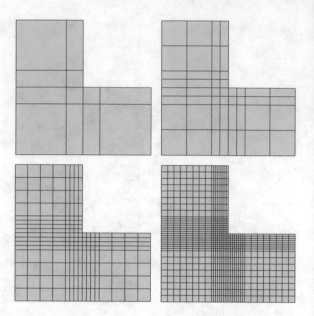

of the fiber. This gives a fiber volume fraction of $\phi_f = 0.004975$ and the bounds $E_c = 3184.075$ and 3013.869.

Third Step:

$$E_c = \frac{\sigma_c}{\varepsilon_c} = 3035.8 \,\text{MPa} \tag{A.66}$$

The fiber volume fraction based on $V_c = V_m = 20000$, i.e. $\phi_f = 0.001$ gives the boundaries $E_c = 3037.00$ and 3002.78. On the other hand, the fiber volume fraction based on $V_c = V_m + V_f = 20020$, i.e. $\phi_f = 0.0009990$ gives the boundaries $E_c = 3036.96$ and 3002.77. The requested meshes are shown in Fig. A.6.

Questions from Chap. 6

6.1.1 Plate Element under Bending Load

See Table A.8.

The EULER- BERNOULLI solution is obtained as:

$$u_z = \frac{-FL^3}{3EI} = -108.0 \ \text{ and } \ |\varphi_y| = \left| \frac{-FL^2}{2EI} \right| = 108.0. \tag{A.67}$$

6.1.2 Simply Supported Plate Element

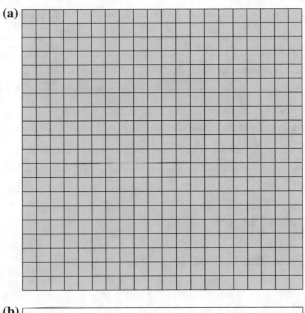

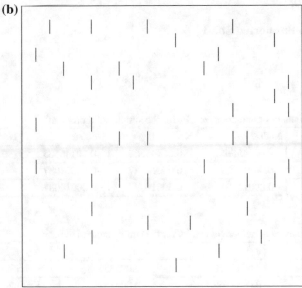

Fig. A.6 Element representation for **a** matrix and **b** fibers

See Table A.9.

The EULER- BERNOULLI solution is obtained as:

$$u_z = \frac{-FL^3}{48EI} = -3.375.$$

(A.68)

Table A.8 Summary of numerical results for the cantilever plate bending problem

Quantity	Node 2	Node 3
u_z	−104.712	−104.712
φ_y	105.744	105.744
φ_x	−6.88142	6.88142

Table A.9 Summary of numerical results for the simply supported plate bending problem

Quantity	(0.75, 0)	(0.75, 0.5)	(0.75, 1.0)
u_z	−3.18921	−3.56747	−3.18921

Questions from Chap. 7

7.2.1 Solid Under Tensile Load

See Table A.10.

7.2.2 Simply Supported Solid

See Table A.11.

Table A.10 Summary of numerical results for the simple solid problem

Quantity	Node 2	Node 3	Node 6	Node 7
u_X	0.00145565	0.00145565	0.00145565	0.00145565
u_Y	0.00014087	0.00014087	−0.00014087	−0.00014087
u_Z	−0.00014087	0.00014087	−0.00014087	0.00014087

Table A.11 Summary of numerical results for the simply supported beam

Node	u_X	u_Y	u_Z
2	0.0017768	−0.00433298	-8.56874×10^{-5}
3	0.0017768	−0.00433298	8.56874×10^{-5}
6	0.00177964	−0.00466403	0.00014218
7	0.00177964	−0.00466403	−0.00014218

Questions from Chap. 8

8.2.1 Tensile Sample with Ideal-Plastic Material Behavior

See Table A.12.

8.2.2 Tensile Sample with Linear Hardening

See Tables A.13 and A.14.

Table A.12 Summary of the numerical values for one element in the case of ideal plasticity (10 increments; $\Delta u_2 = 0.8$ mm)

inc –	u_2 mm	σ MPa	ε 10^{-2}	$\varepsilon^{\mathrm{pl}}$ 10^{-2}	cF_R kN
1	0.8	140	0.2	0.0	14
2	1.6	280	0.4	0.0	28
3	2.4	350	0.6	0.1	35
4	3.2	350	0.8	0.3	35
5	4.0	350	1.0	0.5	35
6	4.8	350	1.2	0.7	35
7	5.6	350	1.4	0.9	35
8	6.4	350	1.6	1.1	35
9	7.2	350	1.8	1.3	35
10	8.0	350	2.0	1.5	35

Table A.13 Numerical values for one element in the case of linear hardening (10 increments; $\Delta u_2 = 0.8$ mm)

inc –	u_2 mm	σ MPa	ε 10^{-2}	$\varepsilon^{\mathrm{pl}}$ 10^{-2}	F_R kN
1	0.8	140	0.2	0.0	14
2	1.6	280	0.4	0.0	28
3	2.4	356.363	0.6	0.0909	35.636
4	3.2	369.091	0.8	0.2727	36.909
5	4.0	381.818	1.0	0.4545	38.182
6	4.8	394.545	1.2	0.6364	39.455
7	5.6	407.272	1.4	0.8182	40.727
8	6.4	419.999	1.6	1.0000	42.000
9	7.2	432.726	1.8	1.1818	43.273
10	8.0	445.453	2.0	1.3636	44.545

Table A.14 Numerical values for one element in the case of linear hardening (10 increments; $\Delta F_2 = 1 \times 10^4$ N)

inc –	u_2 mm	σ MPa	ε 10^{-2}	ε^{pl} 10^{-2}	F_{R} kN
1	0.5714	100	0.1429	0.0	0
2	1.1429	200	0.2857	0.0	0
3	1.7143	300	0.4286	0.0	0
4	3.4007	372.284	0.8501	0.3183	−2771.65
5	11.4288	500	2.8572	2.1429	0
6	17.7146	600	4.4287	3.5715	0
7	24.0004	700	6.0001	5.0001	0
8	30.2863	800	7.5716	6.4287	0
9	36.5721	900	9.1430	7.8573	0
10	42.8580	1000	10.7145	9.2859	0

References

1. Alonso A (2017) Historia del Marc—Un homenaje a Pedro Vicente Marcal. http://www.truegrid.com/pub/History_of_Marc_by_Antonio_Alonso.pdf, http://xyzsa.com/pub/Historia_del_Marc_by_Antonio_Alonso.pdf. Cited 22 May 2017
2. Brown University School of Engineering Press Release: Brown Engineering Alumni H. David Hibbitt Ph.D.'72 and Enrique Lavernia'82 Elected to the National Academy of Engineering (2013)
3. Chen WF, Han DJ (1988) Plasticity for structural engineers. Springer-Verlag, New York
4. Chung DDL (2010) Composite materials: science and applications. Springer-Verlag, London
5. Cook RD, Malkus DS, Plesha ME, Witt RJ (2002) Concepts and applications of finite element analysis. Wiley, New York
6. David Hibbitt–Press Releases Rhode Island Science and Technology Advisory Council (2015). http://stac.ri.gov/about/council/david-hibbitt/. Cited 13 Nov 2015
7. Fish J, Belytschko T (2007) A first course in finite elements. Wiley, Chichester
8. Hulst E (1982) An overview of the MARC general purpose finite element program. In: Brebbia CA (ed) Finite element systems: a handbook, 2nd edn. Springer, Berlin
9. Javanbakht Z, Öchsner A (2017) Advanced finite element simulation with MSC Marc: application of user subroutines. Springer, Cham
10. Javanbakht Z, Öchsner A (2018) Computational statics revision course. Springer, Cham
11. MacNeal Group Press Releases: Icons of CAE Technology Join Forces to Accelerate Product Design. http://www.macnealgroup.com/www/aboutus/press_12_16_05.htm, http://www.zoominfo.com/p/Pedro-Marcal/89295178. Cited 13 Nov 2015
12. MSC Software Corporation (2014) Marc volume B: element library. MSC Software Corporation, Newport Beach
13. MSC Software: Simulating Reality, Delivering Certainty (2015). http://www.mscsoftware.com/page/msc-software. Cited 13 Nov 2015
14. Öchsner A, Merkel M (2013) One-dimensional finite elements: an introduction to the FE method. Springer, Berlin
15. Öchsner A (2014) Elasto-plasticity of frame structure elements: modelling and simulation of rods and beams. Springer, Berlin
16. Öchsner A (2016) Continuum damage and fracture mechanics. Springer, Singapore
17. Öchsner A (2016) Computational statics and dynamics: an introduction based on the finite element method. Springer, Singapore
18. Öchsner A (2018) A project-based introduction to computational statics. Springer, Cham
19. Pilkey WD (2005) Formulas for stress, strain, and structural matrices. Wiley, Hoboken
20. Marcal PV (2015) Stanford Composites Manufacturing Innovation Center. http://web.stanford.edu/group/composites/program/marcal.html. Cited 13 Nov 2015
21. Purushothama Raj P, Ramasamy V (2012) Strength of materials. Pearson, Chennai

© Springer International Publishing AG 2018
A. Öchsner and M. Öchsner, *A First Introduction to the Finite Element Analysis Program MSC Marc/Mentat*, https://doi.org/10.1007/978-3-319-71915-3

22. Reddy JN (2004) An introduction to nonlinear finite element analysis. Oxford University Press, Oxford

23. Reuters: Hexagon to buy U.S. MSC Software for $834 mln in biggest deal since 2010. http://www.reuters.com/article/hexagon-acquistion-msc-software/hexagon-to-buy-u-s-msc-software-for-834-mln-in-biggest-deal-since-2010-idUSL5N1FN1RE. Cited 22 May 2017

24. SAE International: Formula SAE®. http://fsaeonline.com. Cited 31 August 2016

25. The New York Times Company News: MacNeal-Schwendler to buy MARC Analysis Research (1999). http://www.nytimes.com/1999/05/29/business/company-news-macneal-schwendler-to-buy-marc-analysis-research.html. Cited 13 Nov 2015

26. Timoshenko S (1940) Strength of materials-part I elementary theory and problems. D. Van Nostrand Company, New York

27. Timoshenko S (1947) Strength of materials-part II advanced theory and problems. D. Van Nostrand Company, New York

28. Timoshenko S, Woinowsky-Krieger S (1959) Theory of plates and shells. McGraw-Hill Book Company, New York

29. Timoshenko SP, Goodier JN (1970) Theory of elasticity. McGraw-Hill, New York

30. Weisberg DE (2008) The engineering design revolution: the people, companies and computer systems that changed forever the practice of engineering. http://www.cadhistory.net/. Cited 13 Nov 2015

31. Zienkiewicz OC, Taylor RL (2000) The finite element method. Vol. 1: The basis. Butterworth-Heinemann, Oxford

32. Zienkiewicz OC, Taylor RL (2000) The finite element method. Vol. 2: Solid mechanics. Butterworth-Heinemann, Oxford

Index

© Springer International Publishing AG 2018
A. Öchsner and M. Öchsner, *A First Introduction to the Finite Element Analysis
Program MSC Marc/Mentat*, https://doi.org/10.1007/978-3-319-71915-3

Printed in the United States
By Bookmasters